Construction Depth Practice Exams
for the Civil PE Exam

Second Edition

Beth Lin Hartmann, PE

Professional Publications, Inc. • Belmont, California

Benefit by Registering This Book with PPI

- Get book updates and corrections.
- Hear the latest exam news.
- Obtain exclusive exam tips and strategies.
- Receive special discounts.

Register your book at **ppi2pass.com/register**.

Report Errors and View Corrections for This Book

PPI is grateful to every reader who notifies us of a possible error. Your feedback allows us to improve the quality and accuracy of our products. You can report errata and view corrections at **ppi2pass.com/errata**.

CONSTRUCTION DEPTH PRACTICE EXAMS FOR THE CIVIL PE EXAM

Second Edition

Current printing of this edition: 1

Printing History

edition number	printing number	update
1	1	New book.
2	1	New edition. Codes update. Copyright update.

Copyright © 2015, Professional Publications, Inc. All rights reserved.

All content is copyrighted by Professional Publications, Inc. (PPI). No part, either text or image, may be used for any purpose other than personal use. Reproduction, modification, storage in a retrieval system or retransmission, in any form or by any means, electronic, mechanical, or otherwise, for reasons other than personal use, without prior written permission from the publisher is strictly prohibited. For written permission, contact PPI at permissions@ppi2pass.com.

Printed in the United States of America.

PPI
1250 Fifth Avenue
Belmont, CA 94002
(650) 593-9119
ppi2pass.com

ISBN: 978-1-59126-484-2

Library of Congress Control Number: 2014951276

Table of Contents

PREFACE v

ACKNOWLEDGMENTS vii

CODES AND REFERENCES ix

INTRODUCTION xi

NOMENCLATURE xiii

PRACTICE EXAM 1 1

PRACTICE EXAM 2 25

PRACTICE EXAM 1 ANSWER KEY 49

PRACTICE EXAM 1 SOLUTIONS 51

PRACTICE EXAM 2 ANSWER KEY 69

PRACTICE EXAM 2 SOLUTIONS 71

Preface

When the idea for this book was first conceived, acquisitions editor Katie Throckmorton contacted me to ask if I was interested in writing a book of exam-like problems to assist civil construction examinees. I was thrilled to be asked to author it. My problems would support the *Construction Depth Reference Manual*, which is a companion to Michael R. Lindeburg's *Civil Engineering Reference Manual*. In my circle of civil engineering friends, everyone knows "The Lindeburg Book." Although there are many books that we individually find helpful and continually reference, the *Civil Engineering Reference Manual* is the book that we all know and love. I felt so lucky to have been contacted by Katie, and was excited to get to work on writing a book that I hoped would prove just as helpful to examinees as the *Civil Engineering Reference Manual* was to me.

In writing this book's first edition, I immediately set out to review the NCEES specifications for the civil PE exam. I wanted to be as accurate and thorough as I could be so that this book would provide the best possible practice for the exam. I not only returned to the basics of civil and construction engineering, but also gave careful consideration to what civil engineers in the construction discipline should know after four years in the industry. Success on the exam requires both types of knowledge, so I felt confident that drawing on each would be the best approach to a complete review. I was also meticulous in compiling a list of reference materials examinees should bring to the exam.

For this second edition, I have updated the Codes and References section and the Introduction to keep this book current with the latest NCEES civil PE construction depth exam specifications and adopted codes. I have also improved and revised a number of problems based on the code changes and errata submissions from readers.

I hope you will find this book useful in your preparation for the PE exam. Good luck to you as you continue your engineering career.

Beth Lin Hartmann, PE

Acknowledgments

The initial draft of this book could not have been written without the support of five people. Jay Mathes and Ryan Sievers offered incredible assistance and support in crafting and perfecting the problems in this book. U.S. Navy Civil Engineer Corps Officers LT Amy Yoon Honek and LT Sean Hughes, PE, reviewed my completed manuscript and provided excellent feedback for improvements to the book as a whole. Thanks also go to Thomas Schreffler, PE, who performed an amazingly thorough technical review. His comments helped to bring my manuscript to a new level of accuracy and completeness. For this second edition, I'd like to thank three additional people who assisted in the writing and improving of this book's problems: Peter vonQualen; Cliff Plymesser, PE; and Dr. Vern Schaefer, PE.

The product development and implementation team at PPI has also been instrumental to the creation of this second edition. I would like to thank Magnolia Molcan, editorial project manager; Julia Lopez, lead editor; Tyler Hayes, copy editor; Ralph Arcena, EIT, engineering intern; Phil Luna, PE, ME, staff engineer; Cathy Schrott, production services manager; Sarah Hubbard, director of product development and implementation; and Jenny King, associate editor-in-chief, for their hard work and dedication to integrating the changes needed to keep this book current with the exam specifications and adopted codes.

A few others offered their assistance and support in various ways, as well. Tanya Main, my physical therapist, helped me to heal from a shoulder injury caused by poor posture when grading senior design projects. Without her help, it would have been much harder to write for long periods of time and complete this project. Hongtao Dang, my graduate assistant, kept me calm, helped me with my courses, and supported me during the final reviews of this book.

Gayle Lin, my mom, encouraged me to write *Construction Depth Practice Exams* and has always been my number one fan. She is an amazing lady. My dad, George Lin, Jr., DDS. (who passed in 1997) stressed the pursuit of excellence in all endeavors. He showed me that through hard work and passion, all things are possible.

My siblings, David Lin, Rick Lin, Jenny Benson (my identical twin), and Jodie Lin, helped me become the person I am today. Reflecting on my childhood, I am certain our shared experiences influenced my life and career choices. I appreciate our continued loving relationships. My children, Hannah and Nate, inspire me to be a better person. Hannah is the most creative person I know. Nate has a knack for humor and makes me laugh every day.

Finally, I would like to thank my husband, Tim, who continues to support me in everything I do. Without him, I would not have been able to juggle the duties of a wife, mom, and U.S. Navy Civil Engineer Corps Officer. I owe him all my love and gratitude for putting up with the long work hours while I served in the Navy. He continues to appreciate my passion (and tolerate my workaholic nature) as I adjust to my new career as a teacher, consultant, and writer.

In the spirit of Michael R. Lindeburg's *Civil Engineering Reference Manual* Acknowledgements, I'd like to write a secret message to my daughter, Hannah. She enjoys puzzles, so let's see how long this takes her to decode. [Hannah] 645 1j2 1 z215g3v5p 1mw 1n1b3mt 645mt p1w6. 3 1n 1 z2gg2j l2jh4m z2x15h2 4v 645. Lp21h2 x4mg3m52 g4 v4pp4d 645j wj21nh...1mw z2 m3x2 g4 645j zj4gs3j, m1g2.

There are so many other people who cheered me on as I worked on this project. I wish I could name all of you, but I'm afraid I'd get in trouble with PPI for using too much ink. If you're reading this book, you know who you are. Thank you, everyone!

I had a lot of help, but the responsibility for errors is all mine. If you find any mistakes, please submit them using PPI's errata website at **ppi2pass.com/errata**.

Finally, many of my friends know me for one particular saying. I aspire to write a book with this title someday, but in the event I never get the opportunity—don't forget, "It's all about the love."

Beth Lin Hartmann, PE

Codes and References

The information that was used to write this book was based on the exam specifications at the time of publication. However, as with engineering practice itself, the PE examination is not always based on the most current codes or cutting-edge technology. Similarly, codes, standards, and regulations adopted by state and local agencies often lag issuance by several years. It is likely that the codes that are most current, the codes that you use in practice, and the codes that are the basis of your exam will all be different. PPI lists on its website the dates and editions of the codes, standards, and regulations on which NCEES has announced the PE exams are based. It is your responsibility to find out which codes are relevant to your exam. In the meantime, here are the codes that have been incorporated into this edition.

CODES

ACI 318: *Building Code Requirements for Structural Concrete*, 2011, American Concrete Institute, Farmington Hills, MI

ACI 347: *Guide to Formwork for Concrete*, 2004, American Concrete Institute, Farmington Hills, MI (in ACI SP-4, Seventh ed. Appendix)

AISC: *Steel Construction Manual*, Fourteenth ed., 2011, American Institute of Steel Construction, Inc., Chicago, IL

ASCE 7-10: *Minimum Design Loads for Buildings and Other Structures*, 2010, American Society of Civil Engineers, Reston, VA

ASTM C1074-11: *Standard Practice for Estimating Concrete Strength by the Maturity Method*, 2011, ASTM International, West Conshohocken, PA

ASTM D2487-11: *Standard Practice for Classification of Soils for Engineering Purposes (Unified Soil Classification System)*, 2011, ASTM International, West Conshohocken, PA

IBC: *International Building Code*, 2012 ed., International Code Council, Inc., Falls Church, VA

MUTCD: *Manual on Uniform Traffic Control Devices*, 2009 ed., U.S. Department of Transportation, Federal Highway Administration, Washington, DC

OSHA: *Occupational Safety and Health Regulations for the Construction Industry*, 29 CFR Part 1926 (U.S. Federal version), U.S. Department of Labor, Washington, DC

REFERENCES

The following references were used to prepare this book. You may also find them useful references to bring with you to the exam.

Cranes and Derricks, Lawrence K. Shapiro and Jay P. Shapiro. The McGraw-Hill Companies

Developing Your Stormwater Pollution Prevention Plan: A Guide for Construction Sites (EPA-833-R-06-004), 2007, U.S. Environmental Protection Agency, Washington, DC

Formwork for Concrete, M.K. Hurd. American Concrete Institute

Foundation Design, Principles, and Practices, Donald P. Coduto. Prentice Hall

Log of Work-Related Injuries and Illnesses (OSHA's Form 300), Rev. 2004, U.S. Department of Labor, Washington, DC

Project Management for Engineering and Construction, Garold D. Oberlender. The McGraw-Hill Companies

Stormwater Management for Construction Activities: Developing Pollution Prevention Plans and Best Management Practices (EPA-832-R-92-005), U.S. Environmental Protection Agency, Washington, DC

Temporary Structures in Construction, Robert Ratay. The McGraw-Hill Companies

The Architect's Studio Companion: Rules of Thumb for Preliminary Design, Edward Allen and Joseph Iano. John Wiley & Sons, Inc.

Introduction

ABOUT THIS BOOK

Construction Depth Practice Exams includes two exams designed to match the format and specifications of the construction depth section of the civil PE exam. Like the actual exam, the exams in this book contain 40 multiple-choice problems, and each problem takes an average of six minutes to solve. Most of the problems are quantitative, requiring calculations to arrive at the correct option. A few are nonquantitative.

Each of the questions will have four answer options, labeled "A," "B," "C," and "D." If the answer options are numerical, they will be displayed in increasing value. One of the answer options is correct (or, will be "most nearly correct"). The remaining answer options are incorrect and may consist of one or more "logical distractors," the term used by NCEES to designate incorrect options that look correct. Incorrect options represent answers found by making common mistakes. These may be simple mathematical errors, such as failing to square a term in an equation, or more serious errors, such as using the wrong equation.

The solutions in this book are presented step-by-step to help you follow the logical development of the solving approach and to provide examples of how you may want to solve similar problems on the exam.

Some solutions include author commentary that uses the following icons for quick identification.

- *Timesaver:* a technique or approach to reduce problem-solving time
- *Pitfall:* a common pitfall or distractor

Solutions presented for each problem may represent only one of several methods for obtaining the correct answer. Alternative problem-solving methods may also produce correct answers.

ABOUT THE EXAM

The Principles and Practice of Engineering (PE) exam is administered by the National Council of Examiners for Engineering and Surveying (NCEES). The civil PE exam is an eight-hour exam divided into a four-hour morning breadth exam and a four-hour afternoon depth exam. The morning breadth exam consists of 40 multiple-choice problems covering eight areas of general civil engineering knowledge: project planning; means and methods; soil mechanics; structural mechanics; hydraulics and hydrology; geometrics; materials; and site development. As the "breadth" designation implies, morning exam problems are general in nature and wide-ranging in scope. All examinees take the same breadth exam.

For the afternoon depth exam, you must select a depth section from one of the five subdisciplines: construction, geotechnical, structural, transportation, or water resources and environmental. The problems on the afternoon depth exam require more specialized knowledge than those on the morning breadth exam. Topics and the distribution of problems on the construction depth exam are as follows.

- **Earthwork Construction and Layout (6 questions)**

  Excavation and embankment (e.g., cut and fill); borrow pit volumes; site layout and control; earthwork mass diagrams and haul distance; site and subsurface investigations

- **Estimating Quantities and Costs (6 questions)**

  Quantity take-off methods; cost estimating; cost analysis for resource selection; work measurement and productivity

- **Construction Operations and Methods (7 questions)**

  Lifting a rigging; crane stability; dewatering and pumping; equipment operations (e.g., selection, production, economics); deep foundation installation

- **Scheduling (5 questions)**

  Construction sequencing; activity time analysis; critical path method (CPM) network analysis; resource scheduling and leveling; time-cost trade-off

- **Material Quality Control and Production (6 questions)**

  Material properties and testing (e.g., soils, concrete, asphalt); weld and bolt installation; quality control process (QA/QC); concrete proportioning and placement; concrete maturity and early strength evaluation

- **Temporary Structures (7 questions)**

    Construction loads, codes, and standards; formwork; falsework and scaffolding; shoring and reshoring; bracing and anchorage for stability; temporary support of excavation

- **Health and Safety (3 questions)**

    OSHA regulations and hazard identification/abatement; safety management and statistics; work zone and public safety

All problems on the breadth and depth exams are multiple choice. The problem statement includes all information required to solve the problem, followed by four options. Only one of the four options is correct. Each problem is independent, so incorrectly calculating the answer to one problem will not impact subsequent problems.

For further information and tips on how to prepare for the construction depth module of the civil PE exam, consult the *Civil Engineering Reference Manual*, the *Construction Depth Reference Manual*, or PPI's civil PE exam FAQs at **ppi2pass.com/cefaq**.

HOW TO USE THIS BOOK

Prior to taking these practice exams, locate and organize relevant resources and materials as if you are taking the actual exam. Refer to the Codes and References section for guidance on materials. Also, visit **ppi2pass.com/stateboards** for a link to your state's board of engineering, and check for any state-specific restrictions on materials you are allowed to bring to the exam. You should also check NCEES' calculator policy at **ppi2pass.com/calculators** to ensure your calculator can be used on the exam.

The two exams in this book allow you to structure your own exam preparation in the way that is best for you. For example, you might choose to take one exam as a pretest to assess your knowledge and determine the areas in which you need more review, and then take the second after you have completed additional study. Alternatively, you might choose to use one exam as a guide for how to solve different types of problems, reading each problem and solution in kind, and then use the second exam to evaluate what you learned.

Whatever your preferred exam preparation method, these exams will be most useful if you restrict yourself to exam-like conditions when solving the problems. When you are ready to begin an exam, set a timer for four hours. Use only the calculator and references you have gathered for use on the exam. Use the space provided near each problem for your calculations, and mark your answer on the answer sheet.

When you finish taking an exam, check your answers against the answer key to assess your performance. Review the solutions to any problems you answered incorrectly or were unable to answer. Read the author commentaries for tips, and compare your problem-solving approaches against those given in the solutions.

Nomenclature

A	area	ft^2
b	width	ft
B	benefit	$
C	circumference	ft
C	compressive stress	lbf/in^2
C	correction factor	–
C	cost	$
C	runoff coefficient	–
C_c	chemistry coefficient	–
C_w	unit weight coefficient	–
d	distance from center of gravity of load to tipping fulcrum	ft
d_b	distance from center of gravity of boom to tipping fulcrum	ft
d_m	distance from center of gravity of crane to tipping fulcrum	ft
D	dead load	lbf
D	depth	ft
D	distance	ft
D	duration	days
E	earthquake load	lbf
E	modulus of elasticity	$kips/in^2$
F	factor of safety	–
F	force	lbf
F	future worth	$
h	head	ft
h	height	ft
H	horizontal force	lbf
i	effective interest rate	%
i	rainfall intensity	in/hr
k_a	coefficient of active earth pressure	–
K_d	wind directionality	–
K_z	velocity pressure exposure coefficient	–
K_{zt}	topographic factor	–
L	length	ft
L	live load due to occupancy	lbf
n	number of compounding periods in life of asset	–
N	number	–
N	bearing capacity factor	–
p	pressure	lbf/ft^2
P	load	lbf
P	power	hp
P	present worth	$
P	soil loading productivity	yd^3/hr
P	resistance	lbf
q	bearing capacity	lbf/ft^2
q_z	velocity pressure	lbf/ft^2
Q	capacity	lbf
Q	flow rate	gal/min
Q	quantity	various
Q	rate of runoff	ft^3/sec
R	nominal load due to initial rainwater or ice, exclusive of the ponding contribution	lbf
R	rate of placement	ft/hr
R	reaction	lbf
S	snow load	lbf
S	strength	lbf/pile
S_f	skin friction resistance	lbf/ft^2
t	thickness	in
T	temperature	°F
T	tension	lbf
v	velocity	mph
V	volume	yd^3
w	load per unit length	lbf/ft
w	width	ft
w-c	gravimetric water-cement ratio	–
W	weight	lbf
W	wind load	lbf

Symbols

δ	deformation	in
η	pump efficiency	–
γ	unit weight	lbf/ft^3
ϕ	internal angle of friction	deg
ρ	density	lbm/ft^3

Subscripts

a	allowable, annual, or axial
adj	adjusted
ave	average
A	added (by pump)
b	beam, bonds, or boom
c	cement, column, crane, or cut
ca	coarse aggregate
d	downward, drainage, or dry
e	end
elev	elevation
E–W	running east to west
f	fill, floor, footing, fulcrum, or framing
fa	fine aggregate
h	horizontal
j	joist
l	labor, left, or load
L	live
m	masonry or material
max	maximum
n	nails
N–S	running north to south
o	original
p	penetration, piles, profit, or pump

pc	pile cap
r	right or roof
req	required
s	soil, studs, or side-friction
t	toe-bearing or total
td	total dynamic
u	ultimate (factored)
VE	value engineering
w	walls, water, water table, window, or wood

Practice Exam 1

In accordance with the rules established by your state, you may use textbooks, handbooks, bound reference materials, and any approved battery- or solar-powered, silent calculator to work this examination. However, no blank papers, writing tablets, unbound scratch paper, or loose notes are permitted. Sufficient room for scratch work is provided in the Examination Booklet.

You are not permitted to share or exchange materials with other examinees. However, the books and other resources used in this afternoon session do not have to be the same as were used in the morning session.

You will have four hours in which to work this session of the examination. Your score will be determined by the number of questions that you answer correctly. There is a total of 40 questions. All 40 questions must be worked correctly in order to receive full credit on the exam. There are no optional questions. Each question is worth 1 point. The maximum possible score for this section of the examination is 40 points.

Partial credit is not available. No credit will be given for methodology, assumptions, or work written in your Examination Booklet.

Record all of your answers on the Answer Sheet. No credit will be given for answers marked in the Examination Booklet. Mark your answers with the official examination pencil provided to you. Marks must be dark and must completely fill the bubbles. Record only one answer per question. If you mark more than one answer, you will not receive credit for the question. If you change an answer, be sure the old bubble is erased completely; incomplete erasures may be misinterpreted as answers.

If you finish early, check your work and make sure that you have followed all instructions. After checking your answers, you may turn in your Examination Booklet and Answer Sheet and leave the examination room. Once you leave, you will not be permitted to return to work or change your answers.

When permission has been given by your proctor, break the seal on the Examination Booklet. Check that all pages are present and legible. If any part of your Examination Booklet is missing, your proctor will issue you a new Booklet.

WAIT FOR PERMISSION TO BEGIN

Name: _____
        Last   First   Middle Initial

Examinee number: _____

Examination Booklet number: _____

Principles and Practice of Engineering Examination

Afternoon Session
Practice Exam 1

Practice Exam 1 Answer Sheet

Name: _____
      Last   First   Middle Initial

Date: _____

1. Ⓐ Ⓑ Ⓒ Ⓓ	11. Ⓐ Ⓑ Ⓒ Ⓓ	21. Ⓐ Ⓑ Ⓒ Ⓓ	31. Ⓐ Ⓑ Ⓒ Ⓓ
2. Ⓐ Ⓑ Ⓒ Ⓓ	12. Ⓐ Ⓑ Ⓒ Ⓓ	22. Ⓐ Ⓑ Ⓒ Ⓓ	32. Ⓐ Ⓑ Ⓒ Ⓓ
3. Ⓐ Ⓑ Ⓒ Ⓓ	13. Ⓐ Ⓑ Ⓒ Ⓓ	23. Ⓐ Ⓑ Ⓒ Ⓓ	33. Ⓐ Ⓑ Ⓒ Ⓓ
4. Ⓐ Ⓑ Ⓒ Ⓓ	14. Ⓐ Ⓑ Ⓒ Ⓓ	24. Ⓐ Ⓑ Ⓒ Ⓓ	34. Ⓐ Ⓑ Ⓒ Ⓓ
5. Ⓐ Ⓑ Ⓒ Ⓓ	15. Ⓐ Ⓑ Ⓒ Ⓓ	25. Ⓐ Ⓑ Ⓒ Ⓓ	35. Ⓐ Ⓑ Ⓒ Ⓓ
6. Ⓐ Ⓑ Ⓒ Ⓓ	16. Ⓐ Ⓑ Ⓒ Ⓓ	26. Ⓐ Ⓑ Ⓒ Ⓓ	36. Ⓐ Ⓑ Ⓒ Ⓓ
7. Ⓐ Ⓑ Ⓒ Ⓓ	17. Ⓐ Ⓑ Ⓒ Ⓓ	27. Ⓐ Ⓑ Ⓒ Ⓓ	37. Ⓐ Ⓑ Ⓒ Ⓓ
8. Ⓐ Ⓑ Ⓒ Ⓓ	18. Ⓐ Ⓑ Ⓒ Ⓓ	28. Ⓐ Ⓑ Ⓒ Ⓓ	38. Ⓐ Ⓑ Ⓒ Ⓓ
9. Ⓐ Ⓑ Ⓒ Ⓓ	19. Ⓐ Ⓑ Ⓒ Ⓓ	29. Ⓐ Ⓑ Ⓒ Ⓓ	39. Ⓐ Ⓑ Ⓒ Ⓓ
10. Ⓐ Ⓑ Ⓒ Ⓓ	20. Ⓐ Ⓑ Ⓒ Ⓓ	30. Ⓐ Ⓑ Ⓒ Ⓓ	40. Ⓐ Ⓑ Ⓒ Ⓓ

Practice Exam 1

1. The grade line of a 200 ft wide by 1000 ft long project site is shown. The existing grade line is consistent over the entire project.

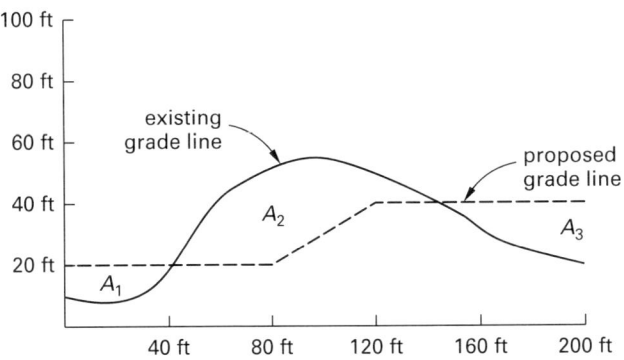

The project site is divided into three areas, A_1, A_2, and A_3. What are most nearly the total volumes of cut and fill, respectively, required to achieve the proposed grade?

(A) 41,000 yd³ cut and 69,000 yd³ fill

(B) 69,000 yd³ cut and 41,000 yd³ fill

(C) 1,100,000 yd³ cut and 1,900,000 yd³ fill

(D) 1,900,000 yd³ cut and 1,100,000 yd³ fill

2. A jobsite with the three cross sections shown is being leveled.

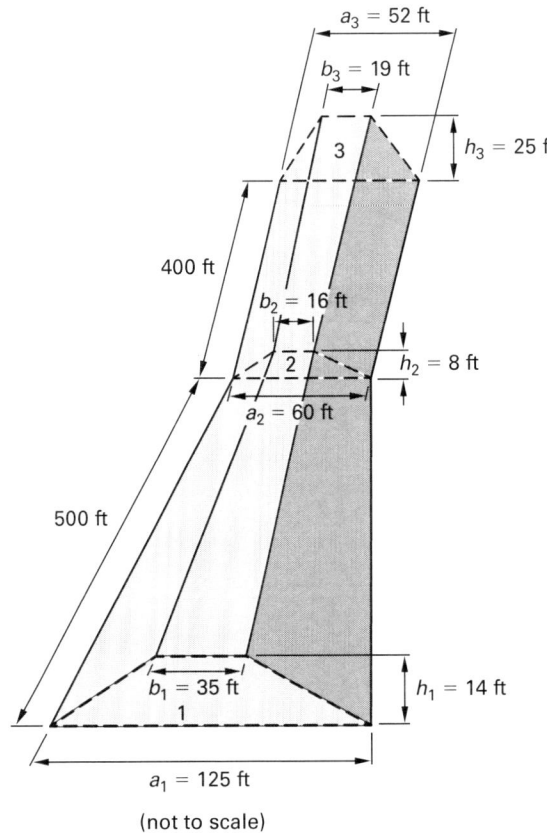

(not to scale)

The volume of soil to be excavated from this site is most nearly

(A) 9000 yd³

(B) 13,000 yd³

(C) 22,000 yd³

(D) 66,000 yd³

3. According to *Occupational Safety and Health Regulations for the Construction Industry* (OSHA), which statement regarding construction site layout is true?

(A) Construction site layout is the responsibility of the subcontractor performing steel erection and/or other structural component installation.

(B) The controlling contractor must ensure all roads immediately outside of the construction site are adequate for safe delivery and movement of equipment.

(C) The controlling contractor is not required to ensure that methods of pedestrian and vehicular control are provided and maintained.

(D) The controlling contractor must ensure adequate space for safe storage of materials.

4. The proposed grade line and existing grade line for a highway project are shown. The highway is 24 ft wide (two 12 ft lanes), and each station is 100 ft long. Any soil cut on-site can be reused on-site. Assume volume and density remain unchanged during soil moving operations.

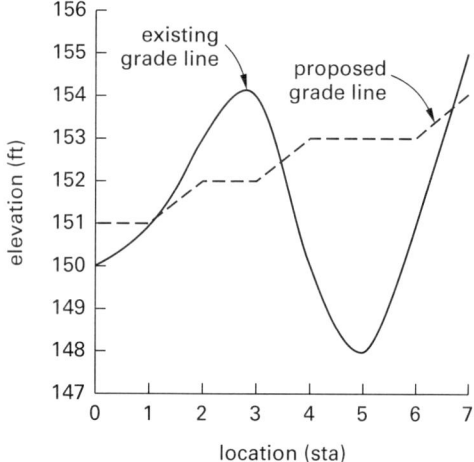

What is most nearly the net earthwork volume of this project?

(A) 540 yd³ fill

(B) 620 yd³ fill

(C) 900 yd³ fill

(D) 1900 yd³ fill

5. 30 ft long precast concrete beams are to be installed in a parking garage. Clear cover on all faces is 3 in. From the cross section shown, what is most nearly the total weight of rebar for each beam?

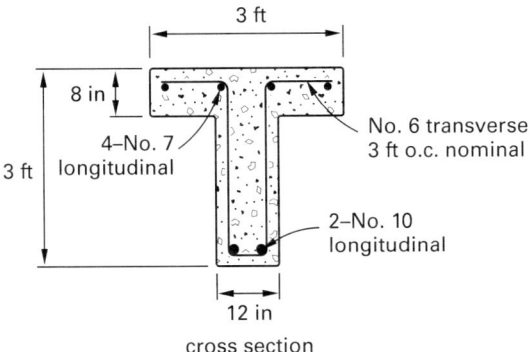

cross section

(A) 608 lbf
(B) 619 lbf
(C) 627 lbf
(D) 652 lbf

6. The plan view of a jobsite is given. The concrete bases of the lightpoles and guardrail supports are designed as shown. All bases must be poured in one day, and the ready-mix trucks from the concrete supplier can hold 10 yd³ of concrete.

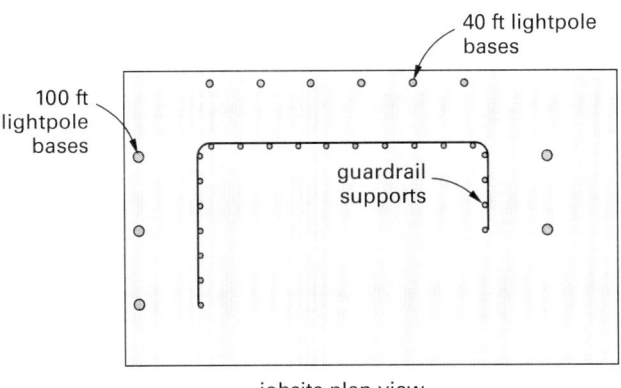

jobsite plan view

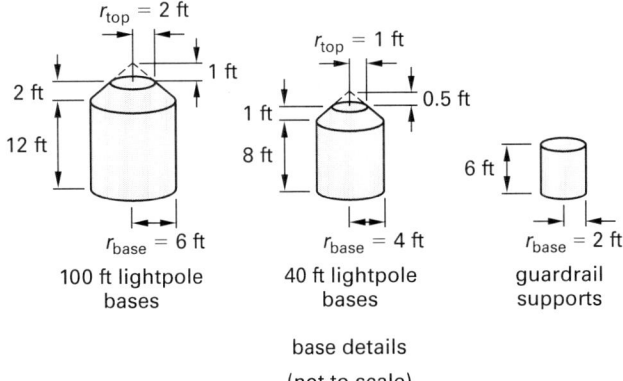

base details
(not to scale)

Given a 5% waste factor, how many trucks of concrete are needed?

(A) 38 trucks
(B) 40 trucks
(C) 43 trucks
(D) 45 trucks

7. The office building shown requires $\frac{5}{8}$ in drywall and all necessary finishing for all exterior walls and the ceiling. The contractor plans to purchase standard 4 ft × 8 ft sheets of drywall. Assume there are no interior walls, and disregard windows, doors, and light fixtures. Costs for labor and materials, including the taxes and delivery costs, are given in the tables.

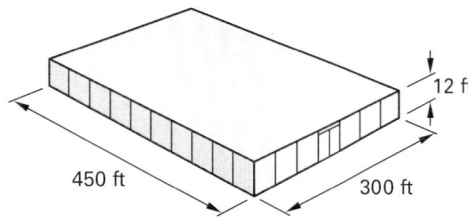

labor rates	
foreman	$33.51/hr
laborer	$23.46/hr

hours required	
drywall	0.8 crew-hr/100 ft²
finishing	1.3 crew-hr/100 ft²

crew composition	
drywall	1 foreman and 2 laborers
finishing	1 foreman and 1 laborer

materials	
drywall and nails	$9.92/sheet
mud and tape	$5.60/100 ft²

Excluding overhead and profit, what is most nearly the total direct cost of the materials and labor needed for the office building?

(A) $169,000

(B) $233,000

(C) $268,000

(D) $314,000

8. The tables give the initial purchase costs and equipment factors of equipment required for a processing plant. The equipment factors are the initial cost multipliers that account for installation costs.

equipment type	cost ($/unit)	quantity
blowers and fans	12,000	1
steam turbine compressors	40,000	3
furnaces	120,000	1
heat exchangers	85,000	2
instrument controls	55,000	3
combustion motors	63,000	1
motor-driven pumps	21,000	1
storage tanks	25,500	2
towers	213,000	1

equipment type	equipment factor
blender	2.0
blowers and fans	2.3
centrifuges	2.0
compressors	
motor-driven	2.3
steam turbine	2.5
ejectors	4.0
furnaces	2.0
heat exchangers	4.8
instrument controls	4.0
motors	
combustion	8.3
electric	7.0
pumps	
motor-driven	7.5
steam turbine	7.0
reactors	5.0
refrigeration unit	2.0
tanks	
process	3.6
storage	2.0
towers	4.0

What is most nearly the direct field cost (DFC) estimate of the project?

(A) $2,600,000

(B) $3,600,000

(C) $3,700,000

(D) $3,800,000

9. A piece of equipment has an initial cost of $35,000. The annual maintenance cost is $1000, and the interest rate is 8%. After five years, the salvage value is $10,000. What is most nearly the equivalent uniform annual cost (EUAC) of the piece of equipment?

(A) $1000

(B) $7000

(C) $8000

(D) $9000

10. Construction on a $100,000 building addition is set to begin in 20 months. The owner of the project wants to invest a lump sum now, and have it accrue enough interest that the total value of the investment at the start of construction will cover the cost. The owner chooses to invest the lump sum in a savings account with a nominal annual interest rate of 10%, compounded monthly. The value of the lump sum the owner must invest is most nearly

(A) $40,000

(B) $59,000

(C) $85,000

(D) $130,000

11. A 58,000 ft² office building is to be constructed. Two competing alternatives are given in the table.

	alternative 1	alternative 2
cost	$10,000,000	$12,000,000
interest rate	10%	10%
operating lifetime	12 yr	20 yr
operating and maintenance costs	$600,000/yr	$450,000/yr
income	$1,900,000/yr	$1,900,000/yr

Which alternative(s) will allow the capital investment in the project to be recovered within the operating lifetime?

(A) both alternative 1 and alternative 2

(B) alternative 1 only

(C) alternative 2 only

(D) neither alternative 1 nor alternative 2

12. A crane holds a 10,000 lbf loading in a stationary position as shown.

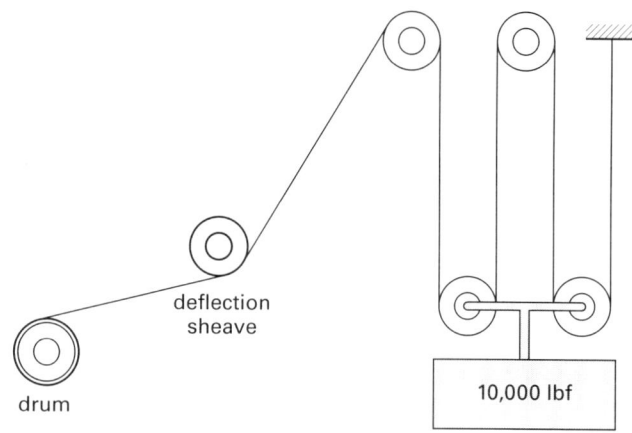

parts of line	lead line factor
2	0.53
3	0.48
4	0.43
5	0.38
6	0.33
7	0.28
8	0.23
9	0.18
10	0.13

What is most nearly the lead line pull on the crane cable?

(A) 3300 lbf

(B) 3800 lbf

(C) 4300 lbf

(D) 4800 lbf

13. Which type of crane would be best suited to the construction of a 30 story, concrete-frame building?

    (A) tower crane

    (B) crawler crane

    (C) lattice boom truck crane

    (D) hydraulic truck crane

14. A construction site must be dewatered at a rate of 200 gal/min for a depth of 35 ft in order to drill caissons. Which of the given water control methods would be best for this job?

    (A) deep well

    (B) wellpoint

    (C) trench

    (D) sump

15. A backhoe with a cycle time of 15 sec is equipped with a 1.5 loose cubic yard (LCY) bucket and can be operated at an efficiency of 45 min/hr. The bucket fill factor is 0.80, and the in-bank correction factor is 0.75. What is most nearly the maximum productivity of the excavator in bank cubic yards per hour (BCY/hr)?

(A) 10 BCY/hr

(B) 160 BCY/hr

(C) 200 BCY/hr

(D) 210 BCY/hr

16. What is the most effective long-term strategy for increasing worker productivity if a project has fallen behind schedule?

(A) have the workers work overtime

(B) provide more efficient equipment

(C) provide additional on-site supervision

(D) all of the above

17. According to Environmental Protection Agency (EPA) standards, which statement regarding the design of a stormwater pollution prevention plan for a construction site is INCORRECT?

(A) Whenever possible, material stockpiles should be placed on paved surfaces to reduce runoff.

(B) All grading activities should be planned so that only a portion of the site is open at any one time.

(C) Sediment controls such as silt fences, sediment barriers, sediment traps, and basins must be in place before initiating activities that could disturb the soil.

(D) Liquid removed through dewatering must be treated before being discharged.

18. Activities C, E, and G shown in the critical path method (CPM) diagram (see *Illustration for Problem 18*) have been delayed and will take twice as long as anticipated to complete. What is the final duration of the project after delays?

(A) 27 days

(B) 30 days

(C) 31 days

(D) 32 days

Illustration for Problem 18

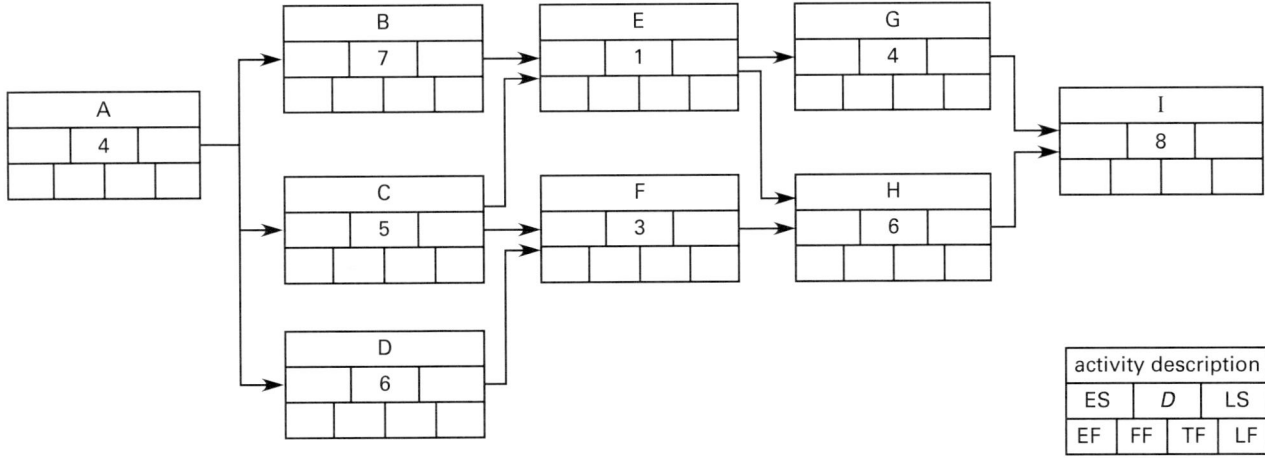

ES: earliest start
EF: earliest finish
LS: latest start
LF: latest finish
FF: free float
TF: total float
D: duration

19. Information regarding a project's activities is given in the table.

activity	duration (days)	predecessors	successors
A	6	none	B, C
B	8	A	D
C	7	A	E
D	3	B	F, G
E	6	C	G, H
F	3	D	I
G	10	D, E	I
H	3	E, F	I
I	1	F, G, H	none

What is the earliest finish of the project?

(A) 21 days

(B) 22 days

(C) 23 days

(D) 30 days

20. Information regarding a project's activities is given in the table.

activity	duration (days)	predecessors	successors
A	5	none	B, C, D
B	2	A	E
C	7	A	F
D	4	A	F, G
E	3	B	H
F	6	C, D	H
G	11	D	I
H	4	E, F	I
I	2	G, H	none

Which activity has the largest amount of free float?

(A) activity A

(B) activity D

(C) activity E

(D) activity G

21. A crew is set to begin priming and painting in five rooms, each with 10,000 ft² of wall area to be worked. The crew can prime 1500 ft²/hr and paint 1000 ft²/hr. All walls need one coat of primer and two coats of paint, and drying time need not be considered. Most nearly how many working days (8 hr/day) will it take to complete the priming and painting?

(A) 11 working days

(B) 14 working days

(C) 17 working days

(D) 21 working days

22. Budgeted costs for a project are given in the table. By the end of month 3, activities A, B, and C have been completed at an actual cost of $9500. Activity D is 75% complete with an actual cost of $9000, and activity E is 50% complete with an actual cost of $8500.

activity	cost ($)	mo. 1	mo. 2	mo. 3	mo. 4	mo. 5	mo. 6	mo. 7
A	2000							
B	1000							
C	4000							
D	14,000							
E	20,000							
F	1900							
G	500							
H	700							

At the end of month 3, most nearly how much has the actual cost of the project strayed from the budgeted cost?

(A) $500 under budget

(B) $500 over budget

(C) $1500 under budget

(D) $1500 over budget

23. Which of the given options is NOT an example of resource leveling?

(A) moving an activity with scheduled float to a later date

(B) moving a crew around a jobsite as needed

(C) waiting to start a critical-path activity until its predecessor has been completed

(D) delaying rebar delivery to free up storage space for bricks and mortar

24. A crew works on a given activity for 8 hr/day at a daily cost of $1200. At the current production rate, it will take 4 weeks (working 5 days per week) to finish the activity. If the crew works four additional hours per day for 3 weeks at an overtime pay rate of 1.5 times the normal pay rate, most nearly how much more than the standard cost will the activity cost to complete?

(A) $3000

(B) $7500

(C) $9000

(D) $12,000

25. An 8 in diameter exploratory boring is drilled through fine sand to a depth of 24 ft. A standard penetration test (SPT) is performed using a U.S. safety hammer and a sampler without a liner. With an SPT N-value of 28 and an energy ratio of 0.60, what is most nearly the corrected SPT N-value for use in field procedures?

(A) 19

(B) 28

(C) 37

(D) 47

26. Which of the given statements is/are true regarding the meaning of the symbol shown?

I. Fillet welds must be used.

II. The dimensions of each weld must be $1/4$ in tall and $1/2$ in wide.

III. Four spot welds spaced 2 in apart must be used to connect $1/4$ in steel plates.

IV. Two 4 in long welds are required along the edge.

V. 2 in long welds must be spaced 4 in apart (centerline-to-centerline).

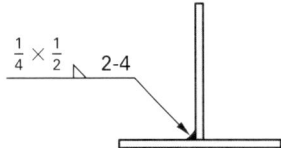

(A) I only

(B) I and III only

(C) I, II, and IV only

(D) I, II, and V only

27. The plan view of a single-story building being designed to meet *International Building Code* (IBC) requirements is shown.

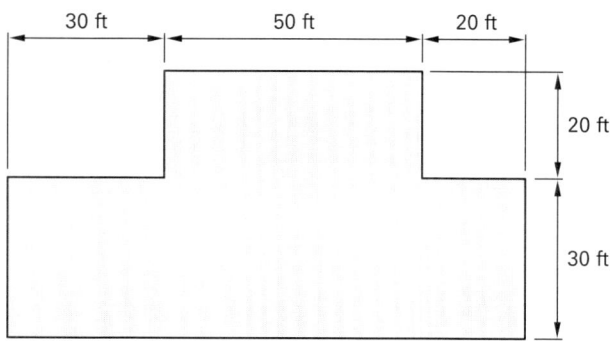

plan view

In order for the building to meet minimum IBC requirements for natural lighting, most nearly how many 3 ft × 4 ft windows must be included in the design?

(A) 18

(B) 22

(C) 27

(D) 32

28. The composition of a concrete mixture is given.

ingredient	quantity	specific gravity	moisture (dry basis from saturated, surface-dry (SSD))
cement	30 sacks (94 lbf/sack)	3.15	–
fine aggregate	6000 lbf	2.65	1.5% deficit (absorption)
coarse aggregate	9500 lbf	2.74	2.0% excess
water	135 gal	–	–
entrained air	3%	–	–

What is most nearly the volumetric concrete yield?

(A) 4.6 yd^3

(B) 4.8 yd^3

(C) 110 yd^3

(D) 130 yd^3

29. A structure has a dead load of 100 lbf/ft^2, a live load of 50 lbf/ft^2, a snow load of 20 lbf/ft^2, and a wind load of 10 lbf/ft^2. What is most nearly the difference between the maximum factored design loads calculated using load and resistance factor design (LRFD) and allowable strength design (ASD)?

(A) 53 lbf/ft^2

(B) 58 lbf/ft^2

(C) 70 lbf/ft^2

(D) 210 lbf/ft^2

30. Scaffolding is being designed with a maximum intended load of 500 lbf. The scaffolding itself will weigh no more than 200 lbf. According to *Occupational Safety and Health Regulations for the Construction Industry* (OSHA), what is most nearly the minimum load that the scaffolding should be designed to support without failure?

(A) 700 lbf

(B) 1200 lbf

(C) 2200 lbf

(D) 2800 lbf

31. According to ASTM C1074, which option is NOT considered a limitation of the maturity method for assessing concrete strength?

(A) The concrete must be maintained at approximately the same conditions as the concrete placed in the field.

(B) The concrete must be maintained in a condition that permits cement hydration.

(C) The effects of early-age concrete temperature on long-term strength are not considered.

(D) The potential strength of the concrete mixture needs to be supplemented by other indicators to achieve an accurate value.

32. 90 mph winds are expected at a jobsite where formwork will be simply supported by braces spaced at 6 ft center-to-center (c.c.) as shown in the illustration.

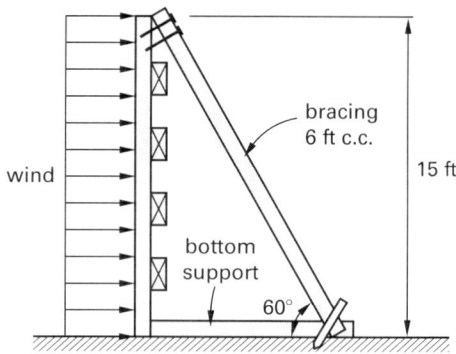

What is most nearly the horizontal component of force in the beam?

(A) 300 lbf

(B) 650 lbf

(C) 935 lbf

(D) 1800 lbf

33. Which method is NOT commonly used to prevent underflow from a sheet pile cofferdam placed on a rock?

(A) increasing the size of dewatering machinery to counteract underflow

(B) leveling the rock surface by removing boulders, cobbles, ridges, and protuberances before the sheet piles are set

(C) placing a thick layer of clay on the outside of the cofferdam cells

(D) placing sacks of concrete on the outside of the cofferdam cells

34. According to *Occupational Safety and Health Regulations for the Construction Industry* (OSHA), when must a fabricated frame scaffold be designed by a registered professional engineer?

(A) when it is 50 ft above the base plate

(B) when it is 75 ft above the base plate

(C) when it is 100 ft above the base plate

(D) when it is 125 ft above the base plate

35. A construction worker must access the roof of an 18 ft tall building with a ladder. What is most nearly the minimum height of ladder that should be used according to *Occupational Safety and Health Regulations for the Construction Industry* (OSHA)?

(A) 18 ft

(B) 22 ft

(C) 24 ft

(D) 26 ft

36. A company has 500 employees who each worked 50 hours each per week for 40 weeks during the construction season last year. Ten recordable injuries or illnesses occurred over the course of the year. What is most nearly the company's Occupational Safety and Health Administration (OSHA) incidence rate?

(A) 1

(B) 2

(C) 4

(D) 10

37. A detention basin bordering a 10 ac unimproved construction site will receive surface rainwater runoff from a watershed area. The water must be pumped 8 ft above the surface of the detention basin to ensure proper dewatering of the site. Runoff coefficients and surface areas for each portion of the watershed area are shown in the illustration. The rainfall intensity is 3.5 in/hr and the pump to be used has an efficiency of 85%.

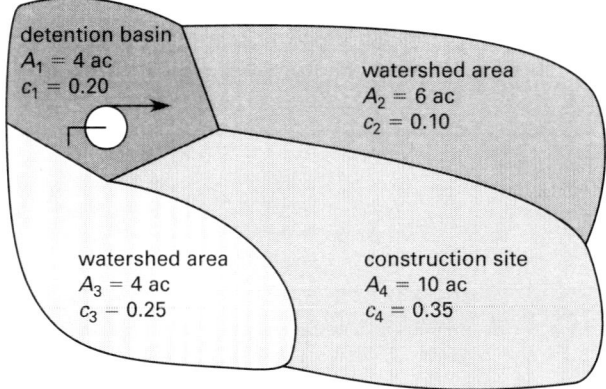

What is most nearly the brake horsepower required to dewater the entire area?

(A) 10 hp

(B) 16 hp

(C) 19 hp

(D) 22 hp

38. A square 6 ft × 6 ft × 2 ft reinforced concrete pile cap transfers vertical loads evenly to four round 18 in diameter reinforced concrete piles that have been previously driven. The pile centers are arranged on a 4.5 ft rectangular grid. The bottom of the pile cap does not contact the soil. The piles penetrate 50 ft into a homogeneous, loose, cohesionless soil. The average effective skin-friction over the length of the piles is 800 lbf/ft^2. The water table is located 5 ft below the surface of the soil. Each pile's end-bearing capacity is considered negligible. Terzaghi's bearing capacity factors are in use.

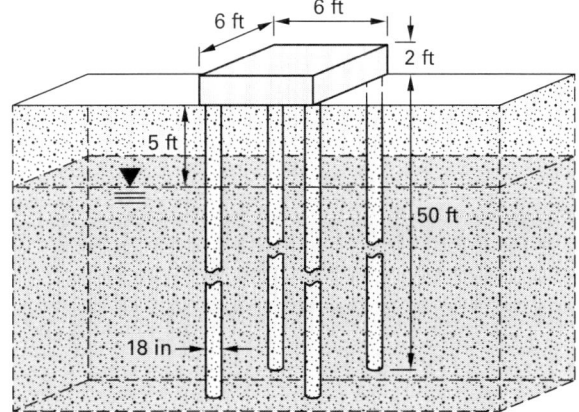

Based on the ultimate capacity for vertical static loading with a factor of safety of 3, what is the design load capacity (in tons of 2000 pounds) of the pile cap?

(A) 90 tons

(B) 100 tons

(C) 110 tons

(D) 130 tons

39. Two 1 in diameter rods hold up a 1000 lbf sign. The modulus of elasticity of the rods is 7500 kips/in^2.

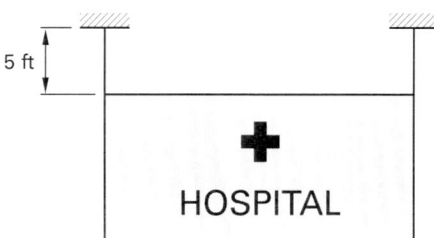

What is most nearly the expected elongation for each loaded rod?

(A) 0.002 in

(B) 0.005 in

(C) 0.010 in

(D) 0.020 in

40. Type II concrete without admixtures is poured into a wall formwork at a rate of 10 ft/hr. The specific weight (unit weight) of the concrete is 145 lbf/ft^3. At the time of pouring, the air temperature is 90°F, and the concrete temperature is 75°F. The wall formwork is 14 ft tall, with vertical studs spaced 9 in horizontally, horizontal wales spaced every 24 in vertically, and ties spaced 32 in. The separation of the two sides of the wall forms is 1 ft. Most nearly, what is the maximum lateral pressure that the formwork will experience when concrete is poured as described?

(A) 900 lbf/ft^2

(B) 1100 lbf/ft^2

(C) 1300 lbf/ft^2

(D) 2000 lbf/ft^2

STOP!

DO NOT CONTINUE!

This concludes the Afternoon Session of the examination. If you finish early, check your work and make sure that you have followed all instructions. After checking your answers, you may turn in your examination booklet and answer sheet and leave the examination room. Once you leave, you will not be permitted to return to work or change your answers.

Practice Exam 2

In accordance with the rules established by your state, you may use textbooks, handbooks, bound reference materials, and any approved battery- or solar-powered, silent calculator to work this examination. However, no blank papers, writing tablets, unbound scratch paper, or loose notes are permitted. Sufficient room for scratch work is provided in the Examination Booklet.

You are not permitted to share or exchange materials with other examinees. However, the books and other resources used in this afternoon session do not have to be the same as were used in the morning session.

You will have four hours in which to work this session of the examination. Your score will be determined by the number of questions that you answer correctly. There is a total of 40 questions. All 40 questions must be worked correctly in order to receive full credit on the exam. There are no optional questions. Each question is worth 1 point. The maximum possible score for this section of the examination is 40 points.

Partial credit is not available. No credit will be given for methodology, assumptions, or work written in your Examination Booklet.

Record all of your answers on the Answer Sheet. No credit will be given for answers marked in the Examination Booklet. Mark your answers with the official examination pencil provided to you. Answers marked in pen may not be graded correctly. Marks must be dark and must completely fill the bubbles. Record only one answer per question. If you mark more than one answer, you will not receive credit for the question. If you change an answer, be sure the old bubble is erased completely; incomplete erasures may be misinterpreted as answers.

If you finish early, check your work and make sure that you have followed all instructions. After checking your answers, you may turn in your Examination Booklet and Answer Sheet and leave the examination room. Once you leave, you will not be permitted to return to work or change your answers.

When permission has been given by your proctor, break the seal on the Examination Booklet. Check that all pages are present and legible. If any part of your Examination Booklet is missing, your proctor will issue you a new Booklet.

WAIT FOR PERMISSION TO BEGIN

Name: _____
       Last   First   Middle Initial

Examinee number: _____

Examination Booklet number: _____

Principles and Practice of Engineering Examination

**Afternoon Session
Practice Exam 2**

Practice Exam 2 Answer Sheet

Name: _____
       Last   First   Middle Initial

Date: _____

41. Ⓐ Ⓑ Ⓒ Ⓓ   51. Ⓐ Ⓑ Ⓒ Ⓓ   61. Ⓐ Ⓑ Ⓒ Ⓓ   71. Ⓐ Ⓑ Ⓒ Ⓓ
42. Ⓐ Ⓑ Ⓒ Ⓓ   52. Ⓐ Ⓑ Ⓒ Ⓓ   62. Ⓐ Ⓑ Ⓒ Ⓓ   72. Ⓐ Ⓑ Ⓒ Ⓓ
43. Ⓐ Ⓑ Ⓒ Ⓓ   53. Ⓐ Ⓑ Ⓒ Ⓓ   63. Ⓐ Ⓑ Ⓒ Ⓓ   73. Ⓐ Ⓑ Ⓒ Ⓓ
44. Ⓐ Ⓑ Ⓒ Ⓓ   54. Ⓐ Ⓑ Ⓒ Ⓓ   64. Ⓐ Ⓑ Ⓒ Ⓓ   74. Ⓐ Ⓑ Ⓒ Ⓓ
45. Ⓐ Ⓑ Ⓒ Ⓓ   55. Ⓐ Ⓑ Ⓒ Ⓓ   65. Ⓐ Ⓑ Ⓒ Ⓓ   75. Ⓐ Ⓑ Ⓒ Ⓓ
46. Ⓐ Ⓑ Ⓒ Ⓓ   56. Ⓐ Ⓑ Ⓒ Ⓓ   66. Ⓐ Ⓑ Ⓒ Ⓓ   76. Ⓐ Ⓑ Ⓒ Ⓓ
47. Ⓐ Ⓑ Ⓒ Ⓓ   57. Ⓐ Ⓑ Ⓒ Ⓓ   67. Ⓐ Ⓑ Ⓒ Ⓓ   77. Ⓐ Ⓑ Ⓒ Ⓓ
48. Ⓐ Ⓑ Ⓒ Ⓓ   58. Ⓐ Ⓑ Ⓒ Ⓓ   68. Ⓐ Ⓑ Ⓒ Ⓓ   78. Ⓐ Ⓑ Ⓒ Ⓓ
49. Ⓐ Ⓑ Ⓒ Ⓓ   59. Ⓐ Ⓑ Ⓒ Ⓓ   69. Ⓐ Ⓑ Ⓒ Ⓓ   79. Ⓐ Ⓑ Ⓒ Ⓓ
50. Ⓐ Ⓑ Ⓒ Ⓓ   60. Ⓐ Ⓑ Ⓒ Ⓓ   70. Ⓐ Ⓑ Ⓒ Ⓓ   80. Ⓐ Ⓑ Ⓒ Ⓓ

Practice Exam 2

41. A jobsite needs 120,000 bank cubic yards (BCY) of soil that will be compacted to a unit weight of 135 lbf/ft^3. Soil with a unit weight of 120 lbf/ft^3 is available for fill from a borrow pit. Most nearly how much fill must be excavated from the borrow pit and brought to the jobsite to complete the earthwork?

(A) 107,000 BCY

(B) 120,000 BCY

(C) 135,000 BCY

(D) 169,000 BCY

42. The existing topography of a borrow pit to be excavated to a depth of 95 ft is shown.

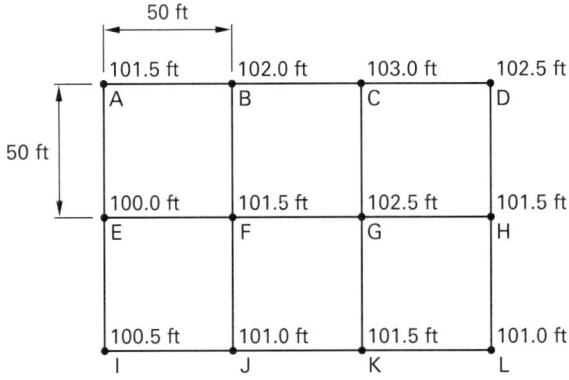

What is most nearly the volume of soil that must be removed from the pit?

(A) 600 yd^3

(B) 3600 yd^3

(C) 45,000 yd^3

(D) 100,000 yd^3

43. Which statement regarding construction site layout is FALSE according to *Occupational Safety and Health Regulations for the Construction Industry* (OSHA)?

(A) Jobsite trailers must be located in the corner opposite the entrance to the site.

(B) The contractor must provide adequate access roads into and through the site.

(C) The contractor must ensure that the construction area is firm and has been properly graded and drained.

(D) The contractor must consider safe storage of materials and safe operation of the erector's equipment when planning the site.

44. Earthwork end areas of cut and fill for six stations of a highway project are given. Assume the soil has a shrinkage factor of 0.2.

station	end area (ft²) cut	fill
0	45	20
1	50	13
2	57	0
3	23	30
4	0	29
5	0	23

What is most nearly the net cut or fill required for these six stations?

(A) 150 yd³ cut

(B) 150 yd³ fill

(C) 300 yd³ cut

(D) 300 yd³ fill

45. The north and west facing walls of the building shown are to receive a brick façade along the bottom 4 ft of the wall. The south and east facing walls are to receive a brick façade for the full height of the building, 16 ft. The face dimensions of the bricks to be used, including the necessary mortar, are 8 in long × 4 in high.

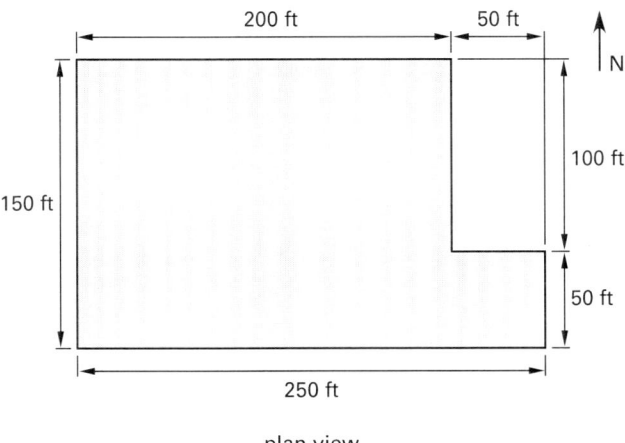

plan view

What is most nearly the number of bricks needed for the brick façade?

(A) 7300 bricks

(B) 29,000 bricks

(C) 36,400 bricks

(D) 58,200 bricks

46. From the quantity takeoff shown, what is most nearly the total weight of steel needed?

	quantity	type	length (ft)
channels	8	MC12 × 40	12
	6	MC8 × 22.8	10
beams	12	W12 × 26	12
	4	W14 × 26	24
	6	W24 × 84	24
	9	W8 × 28	10
columns	14	HSS8 × 3 × 1/4	16
	4	HSS14 × 14 × 3/8	14
	2	HSS16 × 16 × 5/8	14

(A) 10 tons

(B) 12 tons

(C) 17 tons

(D) 19 tons

47. A project is in the bidding stage. The total project cost, not including taxes, bonds, or profit, is $2,750,000, which includes $750,000 for material, $450,000 for labor, and $1,550,000 for subcontracts. A 5% material tax, 18% labor tax, 12% bid bond, and 10% company profit will be included in the bid. Most nearly, what will be the final bid?

(A) $2,820,000

(B) $3,500,000

(C) $3,530,000

(D) $6,920,000

48. A one-story office building with a total of four walls is to be built using wood framing from 2×6 boards. Walls are 10 ft tall including the top and bottom plates. Disregard the additional wood needed for windows, doors, and blocking. Assume that studs are 16 in on center and horizontal framing includes double top plates and single bottom plates for all walls.

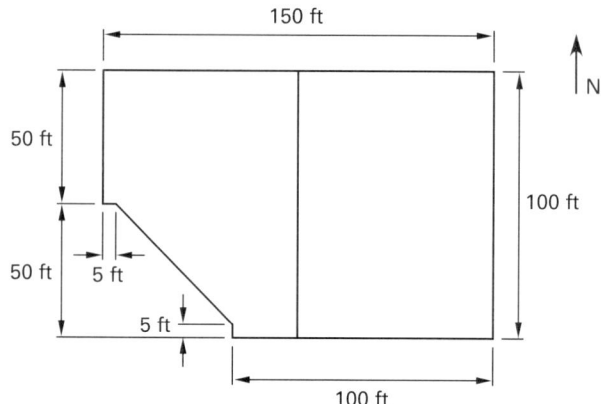

Using a waste factor of 5%, what is most nearly the total length of lumber needed for the interior and exterior framing?

(A) 5800 ft

(B) 5900 ft

(C) 6000 ft

(D) 6200 ft

49. Two competing alternative projects will be evaluated using a desired rate of return of 15%.

	size	cost	annual net profit
project 1	20,000 ft²	$3.25 million	$0.75 million
project 2	50,000 ft²	$6.75 million	$1.9 million

Through value engineering (VE), the cost of project 1 could be reduced by $250,000, and the cost of project 2 could be reduced by $750,000. A project can only be approved if the payback period is five years or shorter. If VE is implemented, which project(s) could be approved?

(A) both project 1 and project 2

(B) project 1 only

(C) project 2 only

(D) neither project 1 nor project 2

50. A proposed facility has an initial cost of $325 million. The capitalized annual cost is $10 million, and the capitalized benefit is $500 million. The facility is expected to have a useful life of 15 years and has an annual depreciation expense of $19 million. The residual value for the facility is estimated at $40 million. What is most nearly the benefit-cost ratio of the project?

(A) 0.60

(B) 0.70

(C) 1.5

(D) 1.7

51. A piece of equipment costs $213,000 to purchase and has a salvage value of $50,000 after 10 years of use. Maintenance costs per year include $600 for tires, $700 for lubricant, and $1300 for fuel. Assume an effective interest rate of 5% compounded yearly. What is most nearly the present worth of the machine?

(A) −$230,000

(B) −$202,000

(C) $230,000

(D) $267,000

52. A crane chart is given in *Illustration for Problem 52* for a crane set up to lift a rooftop onto a 60 ft tall building. The crane's centerline of rotation is 50 ft away from the building, and a 140 ft boom is used at an angle of 60°. The crane is not allowed within 15 ft vertically of the top of the building. What is most nearly the maximum operating radius from the centerline of rotation?

(A) 60 ft

(B) 70 ft

(C) 85 ft

(D) 100 ft

Illustration for Problem 52

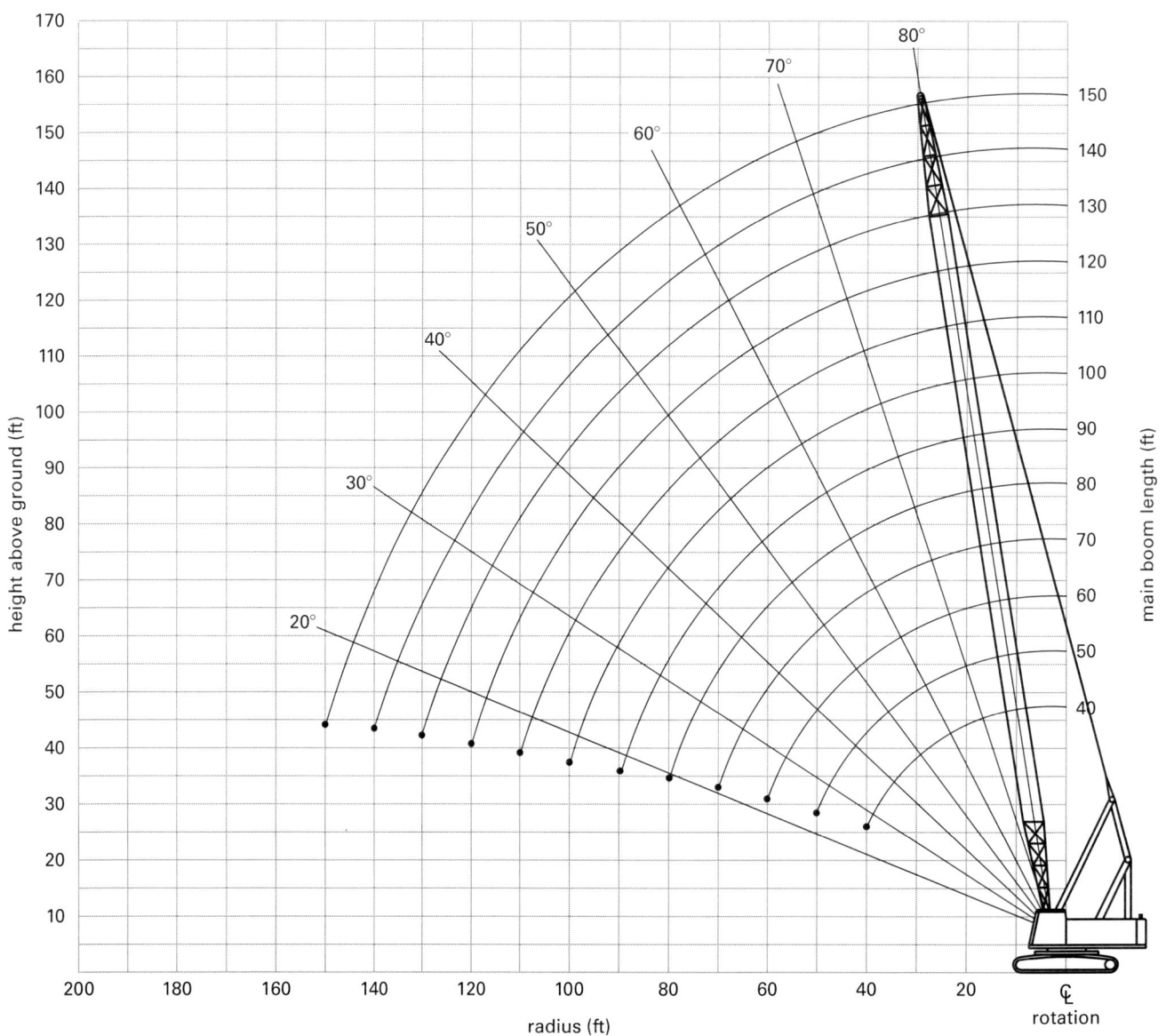

53. The mobile crane shown is unable to set its outriggers due to site constraints. The weight of the crane is 225,000 lbf, the weight of the boom is 36,000 lbf, and the weight of the hook is 1000 lbf.

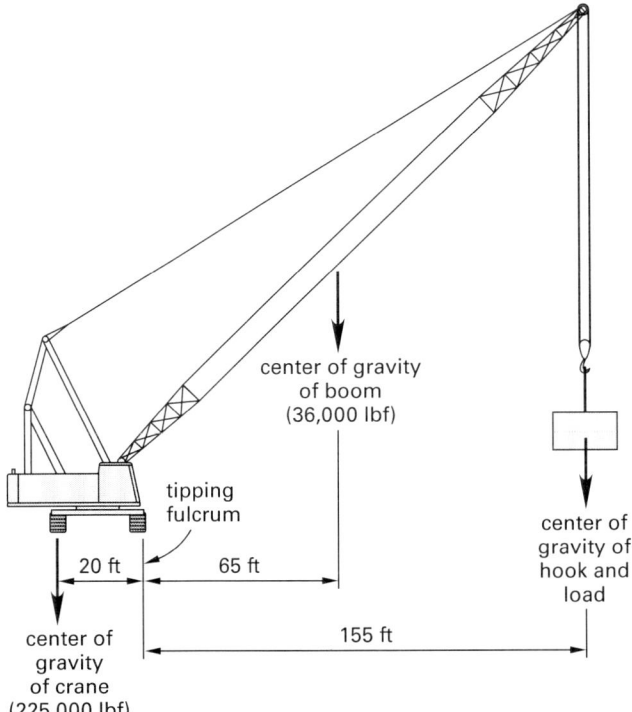

What is most nearly the maximum load the crane can lift without having its outriggers set?

(A) 10,900 lbf

(B) 12,900 lbf

(C) 13,900 lbf

(D) half of the load rating with outriggers set

54. A pump with an expected efficiency of 92% is being sized to pump water from a construction site to a water head of 125 ft at a rate of 300 gal/min. What is most nearly the horsepower needed?

(A) 1.10 hp

(B) 9.50 hp

(C) 10.3 hp

(D) 29.0 hp

55. One excavator and multiple trucks with the given properties are to be used to remove soil. Productivities are given in loose cubic yards per hour (LCY/hr).

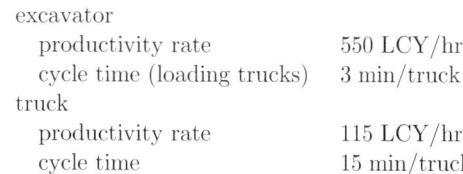

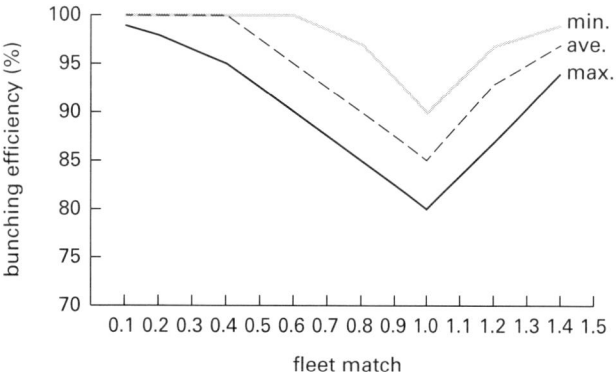

Assuming bunching and queuing are at a maximum, how many trucks should be used to maximize the excavator's productivity?

(A) 3 trucks

(B) 4 trucks

(C) 5 trucks

(D) 6 trucks

56. Three 100 ft × 16 ft masonry walls that will be exposed to the outside are to be constructed. There are 300 ft² of window and door openings that will not receive masonry. Wintery conditions have limited the bricklayers to working in enclosures that can keep an 8 ft tall × 16 ft wide area protected. If the bricklayers can only lay three enclosures per day and the crew's normal productivity is 750 ft²/day, most nearly how much longer will it take to construct the walls in present conditions?

(A) 2 days

(B) 6 days

(C) 10 days

(D) 20 days

57. The current conditions of a 7.5 ac site being planned for redevelopment are given.

surface type	area size	runoff coefficient
grassland	4 ac	0.2
paved parking lot	2 ac	0.9
roof	1.5 ac	0.85

The redevelopment will consist of adding onto the existing building, replacing a portion of the parking lot with porous paving, and adding landscaping elements to the site. After redevelopment, the conditions of the site will be as shown.

surface type	area size	runoff coefficient
grassland	2 ac	0.2
landscaping	1.5 ac	0.05
paved parking lot	1 ac	0.9
porous paving parking	1 ac	0.60
roof	2 ac	0.85

With a rainfall intensity of 2 in/hr, what will be most nearly the runoff prior to and after redevleopment?

(A) 7.35 ft^3/sec prior to redevelopment and 7.75 ft^3/sec after redevelopment

(B) 7.75 ft^3/sec prior to redevelopment and 7.35 ft^3/sec after redevelopment

(C) 9.25 ft^3/sec prior to redevelopment and 7.35 ft^3/sec after redevelopment

(D) 7.35 ft^3/sec prior to redevelopment and 9.25 ft^3/sec after redevelopment

58. Which activities in the project are on the critical path?

activity	duration (days)	predecessors	successors
A	3	none	B, C, D
B	5	A	E
C	7	A	E, F
D	8	A	F, G
E	4	B, C	H
F	6	C, D	H, I
G	9	D	J
H	5	E, F	J
I	6	F	J
J	2	G, H, I	none

(A) A-C-F-H-J

(B) A-C-F-I-J

(C) A-D-G-J

(D) A-D-F-I-J

59. Which scheduling diagram correctly displays the data given in the table?

activity	predecessors	successors
A	–	B
B	A	C, D, E
C	B	G
D	B	F
E	B	F, H
F	D, E	G, I, J
G	C, F	K
H	E	J
I	F	K
J	F, H	M
K	I	L
L	K	N
M	J	N
N	L, M	–

(A) [diagram]

(B) [diagram]

(C) [diagram]

(D) [diagram]

60. Data for a construction project in progress is shown in *Illustration for Problem 60*.

activity	duration	successors	predecessors
A	2	B, D	–
B	2	E	A
C	4	H	D
D	5	C, F	A
E	4	G	B
F	7	J	D, H
G	4	I	E
H	5	F	C
I	8	J	G
J	3	–	I, F

Activity K is added as an additional successor for activities C and E. However, activity K must be completed after activity D, before activity F can occur.

activity	duration	successors	predecessors
C	4	H, K	D
D	5	C	A
E	4	G, K	B
K	6	F	C, D, E

If all other data remains the same, most nearly how many additional days will be needed to complete the project?

(A) 0 days

(B) 1 day

(C) 2 days

(D) 3 days

61. A plumbing crew must install 45 bathroom fixtures in a new building. The crew's normal installation rate is 25 min/fixture, but for this job, there will be a stacking of trades that will cause the crew to work at a productivity efficiency of 75%. Most nearly how many hours will it take the crew to install the fixtures?

(A) 10 hr

(B) 14 hr

(C) 19 hr

(D) 25 hr

Illustration for Problem 60

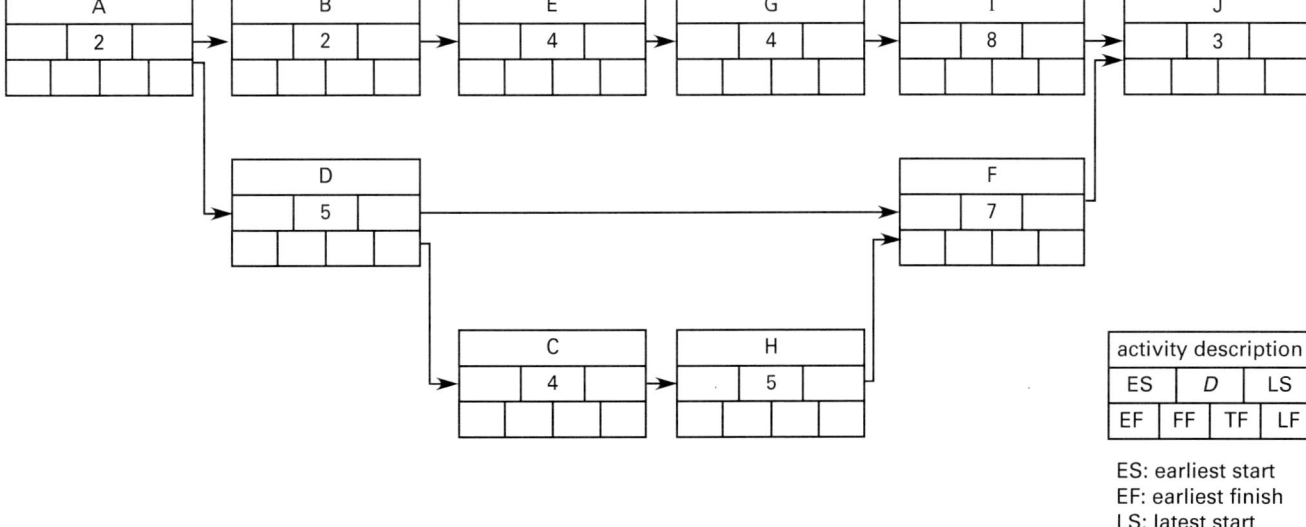

ES: earliest start
EF: earliest finish
LS: latest start
LF: latest finish
FF: free float
TF: total float
D: duration

62. A crew of 3 electricians will install 15,500 linear feet (LF) of conduit in a commercial building at an estimated rate of 120 ft/hr. The work must be completed in 15 working days (8 hr/day) at a budgeted cost of $11,500. The costs associated with the project are given in the table.

	cost
1 apprentice	$20/hr
2 journeymen	$35/hr each
equipment operation	$5/hr

Will the work be completed by the deadline at the budgeted cost?

(A) The work will be completed on time, but over budget.

(B) The work will be completed late, but under budget.

(C) The work will be completed on time and under budget.

(D) The work will be completed late and over budget.

63. The amounts of resources scheduled to be used in four projects are given.

week	project 1 (resources/ wk)	project 2 (resources/ wk)	project 3 (resources/ wk)	project 4 (resources/ wk)
1	1	1	1	3
2	2	4	5	5
3	5	10	5	4
4	6	6	6	5
5	6	6	7	3
6	6	12	12	4
7	8	6	11	5
8	9	4	6	8
9	10	11	10	9
10	8	10	9	3
11	4	3	8	9
12	4	3	4	2
13	2	1	1	1

Which project is considered most level?

(A) project 1

(B) project 2

(C) project 3

(D) project 4

64. 55 rooftop units are to be installed on a large industrial facility. The units can be installed using either a crane or a helicopter. Costs for each method are given.

installation by crane

crane rental	$250/hr
operator	$75/hr
ground crew	$55/hr
rooftop crew	$55/hr

installation by helicopter

helicopter rental	$1000/hr
operator	$300/hr
ground crew	$65/hr
rooftop crew	$65/hr
safety manager	$40/hr

The crane can install 5 units/hr. Assuming any portion of an hour for installation by helicopter will be charged the full hourly rate, what is most nearly the maximum number of hours that operating the helicopter would be more cost effective than operating the crane?

(A) 1 hr

(B) 2 hr

(C) 3 hr

(D) 4 hr

65. A contractor and subcontractor are being considered to dig a 4 ft deep, 300 ft long trench.

	excavator bucket size	excavator cycle time	other considerations
contractor	4 ft wide, 4 yd^3	120 sec/pass	does not have trench box; must slope trench walls at 1:1 slope
subcontractor	3 ft wide, 2 yd^3	55 sec/pass	n.a.

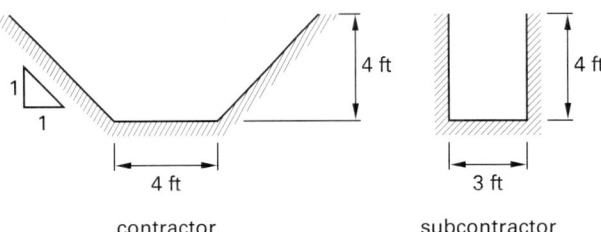

contractor   subcontractor

Who can perform the work in less than 1.5 hr?

(A) contractor

(B) subcontractor

(C) either contractor or subcontractor

(D) neither contractor nor subcontractor

66. A building is constructed according to the plan view shown. All exterior walls are masonry and 25 ft tall. During the inspection process, a problem is found with the quality of the masonry.

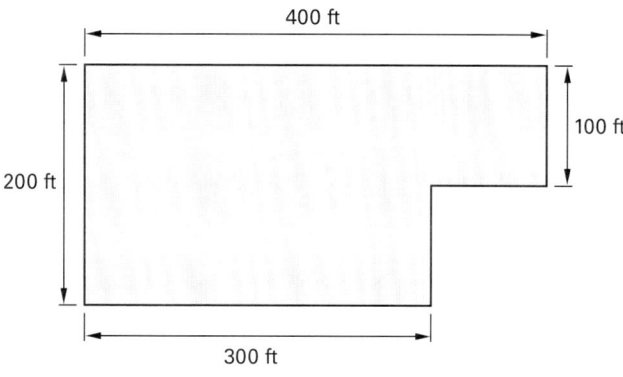

plan view

According to the *International Building Code* (IBC), what is most nearly the minimum number of prism samples necessary to perform a prism test on the masonry?

(A) 9 prisms

(B) 12 prisms

(C) 15 prisms

(D) 18 prisms

67. Which statement is NOT a requirement for an adequately anchored column according to *Occupational Safety and Health Regulations for the Construction Industry* (OSHA)?

(A) Any repairs, replacements, or field modifications must have the written approval of the senior project manager.

(B) It must be able to withstand an eccentric gravity load of 300 lbf with an eccentricity of at least 18 in from its extreme outer face.

(C) It must be set on a level, finished floor adequate to transfer the construction loads.

(D) It must be anchored by a minimum of four anchor bolts.

68. A one-story building with a maximum occupancy of 600 people is being designed. The building will include fire sprinklers. What is the minimum number of 36 in wide openings that the building can have while still meeting or exceeding the minimum egress width for other means of egress (not stairways) according to the *International Building Code* (IBC)?

(A) 2 openings

(B) 3 openings

(C) 4 openings

(D) 5 openings

69. A concrete mixture has the properties shown.

cement	
quantity	10.15 ft^3
specific gravity	3.14
fine aggregate	
quantity	4000 lbf (dry basis)
specific gravity	2.65
moisture excess	+0.5% from SSD
coarse aggregate	
quantity	3500 lbf (dry basis)
specific weight	170 lbf/ft^3
moisture excess	+1.8% from SSD
entrained air	4%

Most nearly how much water is required to obtain a water-cement ratio of 0.45?

(A) 29 gal

(B) 110 gal

(C) 490 gal

(D) 500 gal

70. The beam shown is pin-supported on the left and cable-supported at the center of its span. The cable is attached to a wall above the beam by a bolt screwed into an embedded support.

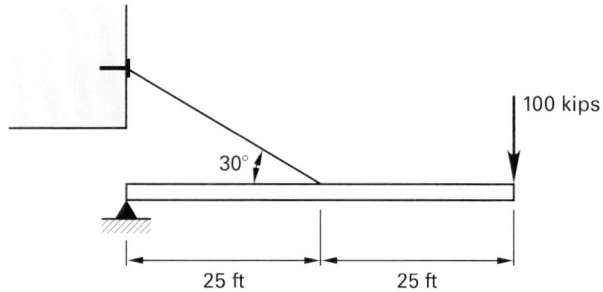

Disregard the weight of the beam. If a force of 100 kips is applied to the end of the beam, what is most nearly the horizontal pull-out force on the bolt?

(A) 150 kips

(B) 250 kips

(C) 350 kips

(D) 400 kips

71. Formwork for a 15 ft high concrete wall is to be constructed from concrete containing Type I cement without any retarders and weighing 130 lbf/ft^3. The concrete will be placed at 70°F. Normal internal vibration will be used with vibrator immersion not exceeding 4 ft. If the maximum design lateral concrete pressure cannot exceed 1200 lbf/ft^2 based on ACI 347, what is most nearly the maximum allowable rate of placement?

(A) 11 ft/hr

(B) 12 ft/hr

(C) 14 ft/hr

(D) 18 ft/hr

72. An 18 ft high concrete retaining wall supports 16 ft of noncohesive soil with an average unit weight of 120 lbf/ft³ and an angle of internal friction of 25° on the right side. On the left side, it supports 5 ft of noncohesive soil with an average unit weight of 130 lbf/ft³ and an angle of internal friction of 35°.

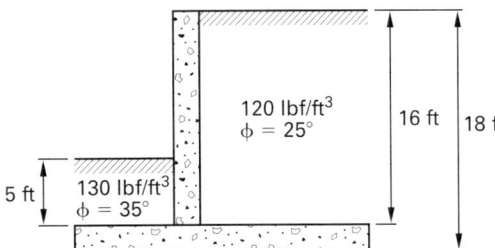

What is most nearly the resultant horizontal force acting on the stem per foot of wall width?

(A) 5800 lbf/ft

(B) 6700 lbf/ft

(C) 12,000 lbf/ft

(D) 17,000 lbf/ft

73. According to ACI 347, which of the statements is NOT one of the assumptions made when shoring and reshoring cast-in-place concrete slabs in multistory buildings with evenly distributed shores?

(A) Shores and reshores are infinitely stiff relative to the slabs.

(B) The topmost slab will deflect no more than twice as much as the slabs below it.

(C) The ground or other base support is rigid, and shores are spaced closely enough to treat shore reactions as distributed loads.

(D) All slabs are identical, and shores and reshores are aligned one-on-one from floor to floor.

74. Which injury is NOT a recordable injury or illness according to *Occupational Safety and Health Regulations for the Construction Industry* (OSHA)?

(A) an injury requiring multiple days away from work to recover

(B) an injury resulting in loss of consciousness

(C) an injury resulting in restricted work or transfer

(D) an injury requiring a tetanus shot

75. Which statement regarding personal protective and lifesaving equipment is NOT true according to *Occupational Safety and Health Regulations for the Construction Industry* (OSHA)?

(A) Lifelines shall be secured to a point capable of supporting a minimum dead weight of 5400 lbf.

(B) It is the employer's responsibility to ensure the sanitation of employee-owned personal protective equipment.

(C) When working over or near water where the danger of drowning exists, employees have the option of taking a swim test or wearing an approved life jacket or buoyant work vest.

(D) Plain cotton is not an acceptable hearing protective device.

76. Which statement regarding scaffolding is FALSE according to *Occupational Safety and Health Regulations for the Construction Industry* (OSHA)?

(A) The stall load of any scaffold hoist shall not exceed 3 times its rated load.

(B) The maximum allowable space between adjacent platform units is 2 in.

(C) Bricklayers' square scaffolds (squares) shall not be more than three tiers high.

(D) Rest platforms shall be provided at least every 35 ft for all supported scaffolds more than 35 ft high.

77. In which of the excavation situations is a registered professional engineer required to design sloping or benching according to *Occupational Safety and Health Regulations for the Construction Industry* (OSHA)?

I. excavating an area with type B soil to a depth of 15 ft for a building foundation

II. excavating an area with type C soil to a depth of 25 ft for building footings

III. excavating an area with type A soil to a depth of 15 ft for building footings

IV. excavating an area with type A soil to a depth of 30 ft for building a foundation

(A) I, II, III, and IV

(B) II, III, and IV only

(C) II and IV only

(D) III and IV only

78. Southern pine 2 × 4s with 19% moisture content during fabrication but more than 19% moisture content in service are to be assembled as shown with 16 penny common nails. One nail will be used on each side. The vertical members are fixed securely at their tops.

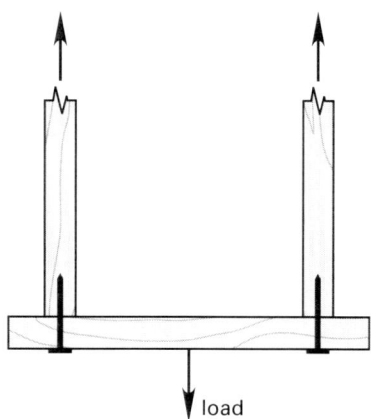

What is most nearly the maximum load that can be applied to the assembly without causing withdrawal failure?

(A) 25 lbf

(B) 50 lbf

(C) 100 lbf

(D) 180 lbf

79. According to *Manual of Uniform Traffic Control Devices* (MUTCD) standards, which statement(s) regarding temporary traffic control is/are true?

I. Flaggers must wear safety apparel visible from a minimum distance of 1200 ft.

II. STOP/SLOW paddles must be a minimum of 18 in wide with letters a minimum of 6 in high.

III. A construction work zone can be flagged without the use of a pilot car if the flaggers can see one another.

IV. If flashing lights are used on the STOP face of a paddle, they must be all white or all yellow.

(A) I, II, III, and IV

(B) I, II, and III only

(C) II only

(D) II and III only

80. A soil sample has a liquid limit (LL) of 45, a plastic limit (PL) of 32, and the characteristics given in the table.

sieve analysis	
sieve no.	% passing
3 in	100
1 1/2 in	100
3/4 in	96
no. 4	58
no. 10	36
no. 20	26
no. 40	–
no. 100	19
no. 200	16

According to the Unified Soil Classification System (USCS), what is the classification of the soil sample?

(A) SM

(B) SC

(C) GC

(D) GM

STOP!

DO NOT CONTINUE!

This concludes the Afternoon Session of the examination. If you finish early, check your work and make sure that you have followed all instructions. After checking your answers, you may turn in your examination booklet and answer sheet and leave the examination room. Once you leave, you will not be permitted to return to work or change your answers.

Practice Exam 1 Answer Key

1. B
2. C
3. D
4. A
5. B
6. D
7. C
8. C
9. C
10. C
11. C
12. C
13. A
14. A
15. B
16. B
17. A
18. D
19. D
20. C
21. C
22. A
23. C
24. B
25. C
26. D
27. C
28. B
29. A
30. C
31. A
32. C
33. A
34. D
35. B
36. B
37. D
38. C
39. B
40. B

Solutions
Practice Exam 1

1. To approximate the cross-sectional volumes of cut and fill required, first identify which areas require cut and which require fill. When the existing grade line is below the proposed grade line, fill is needed. When the existing grade line is above the proposed grade line, cut is needed.

From the illustration, area 1 and area 3 require fill, and area 2 requires cut.

Divide the areas into 10 ft wide rectangles, and use the given graph to estimate the number of square feet of cut or fill required for each area. Use negative numbers to denote fill and positive numbers to denote cut.

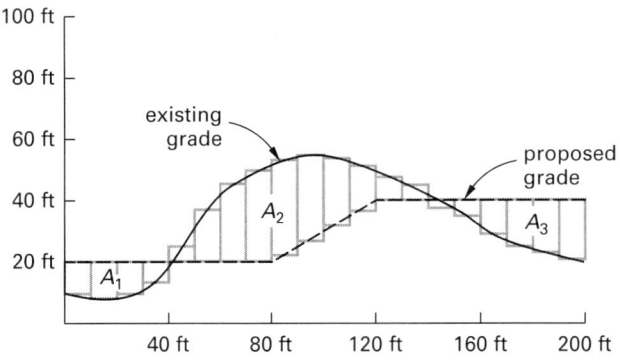

For area 1,

$$A_1 = (10 \text{ ft})(-10 \text{ ft}) + (10 \text{ ft})(-12.5 \text{ ft})$$
$$+ (10 \text{ ft})(-10 \text{ ft}) + (10 \text{ ft})(-7 \text{ ft})$$
$$= -395 \text{ ft}^2 \quad [\text{fill}]$$

For area 2,

$$A_2 = (10 \text{ ft})(5 \text{ ft}) + (10 \text{ ft})(17.5 \text{ ft}) + (10 \text{ ft})(25 \text{ ft})$$
$$+ (10 \text{ ft})(30 \text{ ft}) + (10 \text{ ft})(30 \text{ ft})$$
$$+ (10 \text{ ft})(27.5 \text{ ft}) + (10 \text{ ft})(23 \text{ ft})$$
$$+ (10 \text{ ft})(15 \text{ ft}) + (10 \text{ ft})(7.5 \text{ ft})$$
$$+ (10 \text{ ft})(4.5 \text{ ft})$$
$$= 1850 \text{ ft}^2 \quad [\text{cut}]$$

For area 3,

$$A_3 = (10 \text{ ft})(-2 \text{ ft}) + (10 \text{ ft})(-5 \text{ ft})$$
$$+ (10 \text{ ft})(-11 \text{ ft}) + (10 \text{ ft})(-15 \text{ ft})$$
$$+ (10 \text{ ft})(-17 \text{ ft}) + (10 \text{ ft})(-20 \text{ ft})$$
$$= -700 \text{ ft}^2 \quad [\text{fill}]$$

Calculate the volume of cut and the volume of fill required for the length of 1000 ft.

$$V_c = A_2 L$$
$$= \frac{(1850 \text{ ft}^2)(1000 \text{ ft})}{\left(3 \dfrac{\text{ft}}{\text{yd}}\right)^3}$$
$$= 68{,}519 \text{ yd}^3 \quad (69{,}000 \text{ yd}^3 \text{ cut})$$

$$V_f = (A_1 + A_3)L$$
$$= \frac{(-395 \text{ ft}^2 - 700 \text{ ft}^2)(1000 \text{ ft})}{\left(3 \dfrac{\text{ft}}{\text{yd}}\right)^3}$$
$$= -40{,}556 \text{ yd}^3 \quad (41{,}000 \text{ yd}^3 \text{ fill})$$

The answer is (B).

Author Commentary

🕒 Don't worry about determining the exact areas of cut and fill required. Estimation of quantities utilizing rectangles will yield the correct answer.

💣 When completing mass diagram problems, quantities of cut are represented by positive numbers, while quantities of fill are represented by negative numbers.

2. Determine the volume of soil between section 1 and section 2 using the average end area method.

$$A_1 = \tfrac{1}{2}h_1(a_1 + b_1)$$
$$= \left(\tfrac{1}{2}\right)(14 \text{ ft})(125 \text{ ft} + 35 \text{ ft})$$
$$= 1120 \text{ ft}^2$$
$$A_2 = \tfrac{1}{2}h_2(a_2 + b_2)$$
$$= \left(\tfrac{1}{2}\right)(8 \text{ ft})(60 \text{ ft} + 16 \text{ ft})$$
$$= 304 \text{ ft}^2$$

$$V_{1-2} = L\left(\frac{A_1 + A_2}{2}\right)$$

$$= \frac{(500 \text{ ft})\left(\frac{1120 \text{ ft}^2 + 304 \text{ ft}^2}{2}\right)}{\left(3 \frac{\text{ft}}{\text{yd}}\right)^3}$$

$$= 13{,}185 \text{ yd}^3$$

Determine the volume of soil between section 2 and section 3 using the average end area method.

$$A_2 = \tfrac{1}{2}h_2(a_2 + b_2)$$
$$= \left(\tfrac{1}{2}\right)(8 \text{ ft})(60 \text{ ft} + 16 \text{ ft})$$
$$= 304 \text{ ft}^2$$

$$A_3 = \tfrac{1}{2}h_3(a_3 + b_3)$$
$$= \left(\tfrac{1}{2}\right)(25 \text{ ft})(52 \text{ ft} + 19 \text{ ft})$$
$$= 888 \text{ ft}^2$$

$$V_{2-3} = L\left(\frac{A_2 + A_3}{2}\right)$$

$$= \frac{(400 \text{ ft})\left(\frac{304 \text{ ft}^2 + 888 \text{ ft}^2}{2}\right)}{\left(3 \frac{\text{ft}}{\text{yd}}\right)^3}$$

$$= 8829 \text{ yd}^3$$

Determine the total volume of soil to be excavated.

$$V_t = V_{1-2} + V_{2-3} = 13{,}185 \text{ yd}^3 + 8829 \text{ yd}^3$$
$$= 22{,}014 \text{ yd}^3 \quad (22{,}000 \text{ yd}^3)$$

The answer is (C).

Author Commentary

The average end area method can be used to quickly estimate volume.

💣 Be careful with units when converting between cubic feet and cubic yards.

3. Refer to OSHA Sec. 1926.752(c), which states the following regarding site layout.

Site layout. The controlling contractor shall ensure that the following is provided and maintained.

1926.752(c)(1)

Adequate access roads into and through the site for the safe delivery and movement of derricks, cranes, trucks, other necessary equipment, and the material to be erected and means and methods for pedestrian and vehicular control. Exception: this requirement does not apply to roads outside of the construction site.

1926.752(c)(2)

A firm, properly graded, drained area, readily accessible to the work with adequate space for the safe storage of materials and the safe operation of the erector's equipment.

The answer is (D).

Author Commentary

Site layout and control is an important part of planning construction work, but is often overlooked. Contractors must ensure the safety of their workers and of visitors to the site, ensure the security of their materials, and design a layout that will contribute to maximum productivity.

4. Determine the difference between the existing and proposed elevations at each station.

$$\Delta\text{elev}_{\text{sta}\,0} = 150 \text{ ft} - 151 \text{ ft} = -1 \text{ ft}$$
$$\Delta\text{elev}_{\text{sta}\,1} = 151 \text{ ft} - 151 \text{ ft} = 0 \text{ ft}$$
$$\Delta\text{elev}_{\text{sta}\,2} = 153 \text{ ft} - 152 \text{ ft} = 1 \text{ ft}$$
$$\Delta\text{elev}_{\text{sta}\,3} = 154 \text{ ft} - 152 \text{ ft} = 2 \text{ ft}$$
$$\Delta\text{elev}_{\text{sta}\,4} = 150 \text{ ft} - 153 \text{ ft} = -3 \text{ ft}$$
$$\Delta\text{elev}_{\text{sta}\,5} = 148 \text{ ft} - 153 \text{ ft} = -5 \text{ ft}$$
$$\Delta\text{elev}_{\text{sta}\,6} = 151 \text{ ft} - 153 \text{ ft} = -2 \text{ ft}$$
$$\Delta\text{elev}_{\text{sta}\,7} = 155 \text{ ft} - 154 \text{ ft} = 1 \text{ ft}$$

Find the average change in elevation.

$$\Delta\text{elev}_{\text{ave}} = \frac{\begin{array}{c}\Delta\text{elev}_{\text{sta}\,0} + \Delta\text{elev}_{\text{sta}\,1} + \Delta\text{elev}_{\text{sta}\,2} \\ + \Delta\text{elev}_{\text{sta}\,3} + \Delta\text{elev}_{\text{sta}\,4} + \Delta\text{elev}_{\text{sta}\,5} \\ + \Delta\text{elev}_{\text{sta}\,6} + \Delta\text{elev}_{\text{sta}\,7}\end{array}}{N_{\text{sta}}}$$

$$= \frac{\begin{array}{c}-1 \text{ ft} + 0 \text{ ft} + 1 \text{ ft} + 2 \text{ ft} + (-3 \text{ ft}) \\ + (-5 \text{ ft}) + (-2 \text{ ft}) + 1 \text{ ft}\end{array}}{8}$$

$$= -0.875 \text{ ft} \quad [\text{fill is required}]$$

Multiply the average change in elevation by the length of the project and the width of the highway to find the total volume.

$$V_t = \Delta\text{elev}_{\text{net}} L w$$
$$= \frac{(-0.875 \text{ ft})(700 \text{ ft})(24 \text{ ft})}{\left(3 \frac{\text{ft}}{\text{yd}}\right)^3}$$
$$= -544 \text{ yd}^3 \quad (540 \text{ yd}^3 \text{ fill})$$

The answer is (A).

Author Commentary

Earthwork diagrams are often more complex than this. If soil could not be reused on site, or if reused soil had to be compacted, further calculations would be needed to accurately estimate cut and fill volumes.

💣 When completing mass diagram problems, quantities of cut are represented by positive numbers, while quantities of fill are represented by negative numbers.

5. Calculate the quantity of no. 6 transverse bars.

$$N_{\text{spaces}} = \frac{30 \text{ ft} - \dfrac{3 \text{ in}}{12 \dfrac{\text{in}}{\text{ft}}} - \dfrac{3 \text{ in}}{12 \dfrac{\text{in}}{\text{ft}}}}{3 \text{ ft o.c.}}$$

$$= 9.83 \text{ spaces} \quad (10 \text{ spaces})$$

$$Q_{\text{bars}} = \text{whole number of spaces} + 1$$
$$= 10 + 1$$
$$= 11$$

From the illustration, four no. 7 longitudinal bars are needed, and two no. 10 longitudinal bars are needed.

Determine the length of each size of bar to be used.

For no. 6 transverse,

$$L_{\text{no. 6}} = 1 \text{ ft} + 2.5 \text{ ft} + 0.5 \text{ ft} + 2.5 \text{ ft} + 1 \text{ ft}$$
$$= 7.5 \text{ ft}$$

For no. 7 longitudinal,

$$L_{\text{no. 7}} = 30 \text{ ft} - \dfrac{3 \text{ in}}{12 \dfrac{\text{in}}{\text{ft}}} + \dfrac{3 \text{ in}}{12 \dfrac{\text{in}}{\text{ft}}}$$
$$= 29.5 \text{ ft}$$

For no. 10 longitudinal,

$$L_{\text{no. 10}} = 30 \text{ ft} - \dfrac{3 \text{ in}}{12 \dfrac{\text{in}}{\text{ft}}} + \dfrac{3 \text{ in}}{12 \dfrac{\text{in}}{\text{ft}}}$$
$$= 29.5 \text{ ft}$$

Calculate the total weight of each size of bar. Weights per unit length can be found in the rebar chart in ACI 318 App. E.

$$W_{\text{no. 6}} = QLw$$
$$= (11)(7.5 \text{ ft})\left(1.502 \dfrac{\text{lbf}}{\text{ft}}\right)$$
$$= 123.915 \text{ lbf}$$

$$W_{\text{no. 7}} = QLw$$
$$= (4)(29.5 \text{ ft})\left(2.044 \dfrac{\text{lbf}}{\text{ft}}\right)$$
$$= 241.192 \text{ lbf}$$

$$W_{\text{no. 10}} = QLw$$
$$= (2)(29.5 \text{ ft})\left(4.303 \dfrac{\text{lbf}}{\text{ft}}\right)$$
$$= 253.877 \text{ lbf}$$

$$W_t = W_{\text{no. 6}} + W_{\text{no. 7}} + W_{\text{no. 10}}$$
$$= 123.915 \text{ lbf} + 241.192 \text{ lbf} + 253.877 \text{ lbf}$$
$$= 618.984 \text{ lbf} \quad (619 \text{ lbf})$$

The answer is (B).

Author Commentary

When completing a quantity takeoff, accuracy is important so mistakes are not carried over to subsequent calculations. For example, if a simple quantity takeoff for one beam is used to represent multiple other beams, any mistakes that are made on the simple quantity takeoff will be proliferated in the final quantity takeoff. Also, be sure to multiply the bar lengths by the correct unit weight for each bar size.

🕒 Tab the rebar chart in ACI 318 App. E.

6. Determine the volume of concrete required for the flattened conical tops of the lightpole bases. This volume can be derived by using geometry to calculate the top portion as a full cone and subtracting a smaller cone volume. The smaller cone will use the radius r_{top} for its volume calculation.

$$V = \tfrac{1}{3}\pi r_{\text{base}}^2 h_{\text{full}} - \tfrac{1}{3}\pi r_{\text{top}}^2 h_{\text{partial}}$$

$$V_{\text{top,100 ft}} = \left(\tfrac{1}{3}\right)\pi(6 \text{ ft})^2(2 \text{ ft} + 1 \text{ ft}) - \left(\tfrac{1}{3}\right)\pi(2 \text{ ft})^2(1 \text{ ft})$$
$$= 109 \text{ ft}^3$$

$$V_{\text{top,40 ft}} = \left(\tfrac{1}{3}\right)\pi(4 \text{ ft})^2(1 \text{ ft} + 0.5 \text{ ft}) - \left(\tfrac{1}{3}\right)\pi(1 \text{ ft})^2(0.5 \text{ ft})$$
$$= 25 \text{ ft}^3$$

Determine the volume of concrete required for the cylinders of each base.

$$V_{\text{cyl}} = \pi r_{\text{base}}^2 h_{\text{full}}$$

$$V_{\text{cyl,100 ft}} = \pi(6 \text{ ft})^2(12 \text{ ft})$$
$$= 1357 \text{ ft}^3$$

$$V_{\text{cyl,40 ft}} = \pi(4 \text{ ft})^2(8 \text{ ft})$$
$$= 402 \text{ ft}^3$$
$$V_{\text{cyl,guardrail}} = \pi(2 \text{ ft})^2(6 \text{ ft})$$
$$= 75 \text{ ft}^3$$

Calculate the volume of concrete required for each type of base.

$$V_{\text{cyl,100 ft}} = (V_{\text{top,100 ft}} + V_{\text{cyl,100 ft}})N_{\text{100 ft}}$$
$$= (109 \text{ ft}^3 + 1357 \text{ ft}^3)(5)$$
$$= 7330 \text{ ft}^3$$
$$V_{\text{cyl,40 ft}} = (V_{\text{top,40 ft}} + V_{\text{cyl,40 ft}})N_{\text{40 ft}}$$
$$= (25 \text{ ft}^3 + 402 \text{ ft}^3)(6)$$
$$= 2562 \text{ ft}^3$$
$$V_{\text{cyl,guardrail}} = V_{\text{guardrail}} N_{\text{guardrail}}$$
$$= (75 \text{ ft}^3)(21)$$
$$= 1575 \text{ ft}^3$$

Calculate the total concrete volume.

$$V_{c,t} = V_{t,\text{100 ft}} + V_{t,\text{40 ft}} + V_{t,\text{guardrail}}$$
$$= 7330 \text{ ft}^3 + 2562 \text{ ft}^3 + 1575 \text{ ft}^3$$
$$= 11{,}467 \text{ ft}^3$$

Multiply the combined volume by the waste factor to determine the total amount of concrete wasted.

$$V_{t,\text{waste}} = V_{c,t} \text{WF}$$
$$= (11{,}467 \text{ ft}^3)(0.05)$$
$$= 573.35 \text{ ft}^3$$

The total amount of concrete needed is

$$V_{t,\text{needed}} = V_{c,t} + V_{t,\text{waste}}$$
$$= 11{,}467 \text{ ft}^3 + 573.35 \text{ ft}^3$$
$$= 12{,}040.35 \text{ ft}^3$$

Therefore, the number of trucks required is

$$N_{\text{trucks}} = \frac{12{,}040.35 \text{ ft}^3}{\left(10 \frac{\text{yd}^3}{\text{truck}}\right)\left(3 \frac{\text{ft}}{\text{yd}}\right)^3}$$
$$= 44.59 \text{ trucks} \quad (45 \text{ trucks})$$

The answer is (D).

Author Commentary

It is often difficult to accurately estimate the quantity of concrete needed for a project. The uniformity of earth-formed foundations like the ones in this problem depends entirely on the boring machines and the crew's ability to build the foundations consistently. Generally, waste factors and contingencies are used to account for any inaccuracies.

🕐 This solution can be simplified by calculating the volume of the top portions of the lightpole bases as full cones, rather than as partial cones. The rounded value is still 45 trucks.

7. Determine the area of the office to be drywalled.

$$A_{\text{walls}} = (2)(300 \text{ ft})(12 \text{ ft}) + (2)(450 \text{ ft})(12 \text{ ft})$$
$$= 18{,}000 \text{ ft}^2$$
$$A_{\text{ceiling}} = (300 \text{ ft})(450 \text{ ft})$$
$$= 135{,}000 \text{ ft}^2$$
$$A_t = A_{\text{walls}} + A_{\text{ceiling}}$$
$$= 18{,}000 \text{ ft}^2 + 135{,}000 \text{ ft}^2$$
$$= 153{,}000 \text{ ft}^2$$

Determine the cost of the materials.

For drywall,

$$C_{m,\text{drywall}} = \frac{(153{,}000 \text{ ft}^2)\left(9.92 \frac{\$}{\text{sheet}}\right)}{32 \frac{\text{ft}^2}{\text{sheet}}}$$
$$= \$47{,}430$$

For mud and tape,

$$C_{m,\text{mud and tape}} = (153{,}000 \text{ ft}^2)\left(5.60 \frac{\$}{100 \text{ ft}^2}\right)$$
$$= \$8568$$

The total cost of materials is

$$C_{m,t} = C_{m,\text{drywall}} + C_{m,\text{mud and tape}}$$
$$= \$47{,}430 + \$8568$$
$$= \$55{,}998$$

Determine the cost of labor.

For drywall,

$$C_{l,\text{drywall}} = \left(\left(1 \frac{\text{foreman}}{\text{crew}}\right)\left(33.51 \frac{\$}{\text{hr}}\right) + \left(2 \frac{\text{laborers}}{\text{crew}}\right)\left(23.46 \frac{\$}{\text{hr-laborer}}\right)\right)$$
$$\times \left(0.8 \frac{\text{crew-hr}}{100 \text{ ft}^2}\right)(153{,}000 \text{ ft}^2)$$
$$= \$98{,}446.32$$

For finishing,

$$C_{l,\text{finishing}} = \left(33.51 \ \frac{\$}{\text{hr}} + 23.46 \ \frac{\$}{\text{hr}}\right)$$
$$\times \left(1.3 \ \frac{\text{crew-hr}}{100 \ \text{ft}^2}\right)(153{,}000 \ \text{ft}^2)$$
$$= \$113{,}313.33$$

The total cost of labor is

$$C_{l,t} = C_{l,\text{drywall}} + C_{l,\text{finishing}}$$
$$= \$98{,}446.32 + \$113{,}313.33$$
$$= \$211{,}759.65$$

The total direct cost of materials and labor is

$$C_t = C_{m,t} + C_{l,t}$$
$$= \$55{,}998 + \$211{,}759.65$$
$$= \$267{,}757.65 \quad (\$268{,}000)$$

The answer is (C).

Author Commentary

- Make sure to multiply the hourly rate by the number of workers in the crew for each activity.

8. For each type of equipment, multiply the unit cost by the number of units and the equipment factor.

$$C_{\text{blowers and fans}} = (\$12{,}000)(1)(2.3)$$
$$= \$27{,}600$$
$$C_{\text{steam turbine compressors}} = (\$40{,}000)(3)(2.5)$$
$$= \$300{,}000$$
$$C_{\text{furnaces}} = (\$120{,}000)(1)(2.0)$$
$$= \$240{,}000$$
$$C_{\text{heat exchangers}} = (\$85{,}000)(2)(4.8)$$
$$= \$816{,}000$$
$$C_{\text{instrument controls}} = (\$55{,}000)(3)(4.0)$$
$$= \$660{,}000$$
$$C_{\text{combustion motors}} = (\$63{,}000)(8.3)$$
$$= \$522{,}900$$
$$C_{\text{motor-driven pumps}} = (\$21{,}000)(7.5)$$
$$= \$157{,}500$$
$$C_{\text{storage tanks}} = (\$25{,}500)(2)(2.0)$$
$$= \$102{,}000$$
$$C_{\text{towers}} = (\$213{,}000)(1)(4.0)$$
$$= \$852{,}000$$

The DFC is

$$C_t = C_{\text{blowers and fans}} + C_{\text{steam turbine compressors}}$$
$$+ C_{\text{furnaces}} + C_{\text{heat exchangers}} + C_{\text{instrument controls}}$$
$$+ C_{\text{combustion motors}} + C_{\text{motor-driven pumps}}$$
$$+ C_{\text{storage tanks}} + C_{\text{towers}}$$
$$= \$27{,}600 + \$300{,}000 + \$240{,}000$$
$$+ \$816{,}000 + \$660{,}000 + \$522{,}900$$
$$+ \$157{,}500 + \$102{,}000 + \$852{,}000$$
$$= \$3{,}678{,}000 \quad (\$3{,}700{,}000)$$

The answer is (C).

Author Commentary

Conceptual estimating is often used when owners are trying to budget money for a future construction project. In many industrial plants, the most expensive parts of the facility are major pieces of equipment used in production. A direct field cost estimate allows owners to project the cost of a building based on historical data of the equipment cost.

9. Use a cash flow equivalent factor table to find the cash flow equivalent factor for five years and an 8% interest rate.

$$\text{EUAC} = P(A/P, i\%, n) + \$1000 - S(A/F, i\%, n)$$
$$= (\$35{,}000)(0.2505) + \$1000$$
$$- (\$10{,}000)(0.1705)$$
$$= \$8062.50 \quad (\$8000)$$

The answer is (C).

Author Commentary

- Tab the factor tables in a reference manual for quick access during the exam.
- Make sure that you select the correct factor table and read the correct column and row.

10. Determine the effective monthly interest rate.

$$i_{\text{monthly}} = \frac{i}{12 \ \text{mo}} = \frac{0.10}{12 \ \text{mo}}$$
$$= 0.00833$$

Calculate the present worth.

$$P = F(1+i)^{-n} = \frac{F}{(1+i)^n}$$

$$= \frac{\$100{,}000}{(1+0.0083)^{20}}$$

$$= \$84{,}762.67 \quad (\$85{,}000)$$

The answer is (C).

Author Commentary

Engineering economics studies the time value of money. Money that is saved now can be worth more in the future if it is managed correctly. Knowing how to convert an annual amount to a present worth is very valuable when planning for a project.

11. Use the capital recovery method to determine the economic feasibility of each alternative.

For alternative 1, the net annual income required to recover the capital investment is

$$A_1 = P\left(\frac{i(1+i)^n}{(1+i)^n - 1}\right) + C_{\text{OM}}$$

$$= \left(10{,}000{,}000 \ \frac{\$}{\text{yr}}\right)\left(\frac{(0.10)(1+0.10)^{12}}{(1+0.10)^{12} - 1}\right)$$

$$+ 600{,}000 \ \frac{\$}{\text{yr}}$$

$$= \$2{,}067{,}633/\text{yr} \quad [>\$1{,}900{,}000/\text{yr}]$$

Alternative 1 will not allow capital investment to be recovered.

For alternative 2, the net annual income required to recover the capital investment is

$$A_2 = P\left(\frac{i(1+i)^n}{(1+i)^n - 1}\right) + C_{\text{OM}}$$

$$= \left(12{,}000{,}000 \ \frac{\$}{\text{yr}}\right)\left(\frac{(0.10)(1+0.10)^{20}}{(1+0.10)^{20} - 1}\right)$$

$$+ 450{,}000 \ \frac{\$}{\text{yr}}$$

$$= \$1{,}859{,}515/\text{yr} \quad [<\$1{,}900{,}000/\text{yr}]$$

Alternative 2 will allow capital investment to be recovered.

The answer is (C).

Author Commentary

Value engineering involves evaluating life cycle costs and alternatives. It is important to understand that the least expensive initial investment will not always be the most economical.

- Review engineering economics formulas. Understand and note how each is used and be able to quickly recall the correct formula for each engineering economics problem on the exam.
- When using the capital recovery method, the value obtained is the minimum amount needed to recover the initial investment. If the number is less than the expected revenue, the scenario is economically feasible.

12. Determine the number of parts of line in the rigging configuration.

There are five sheaves. However, one line is a deflection sheave and is not considered when counting parts of line. Four vertical lines are helping to lift the block. Therefore, this configuration is a four-part line. From the given table, the lead line factor for a four-part line is 0.43. Therefore, the lead line pull, or tension, on the crane cable is

$$T = (0.43)(10{,}000 \text{ lbf})$$

$$= 4300 \text{ lbf}$$

The answer is (C).

Author Commentary

In many crane configurations, there are different types of sheaves to consider. Typically, cranes and crane rigging equipment come with charts for efficiency, lead line factor, and load. Each of these must be carefully considered when developing a lifting plan. When determining how many parts are required for a reeving system, both sheave efficiency and lead line factors are used to calculate capacities and tensions within the lifting system.

13. The stability of crawler cranes comes from counterweights near the ground. It would be impractical for a crawler crane to have the necessary weight at the ground level to support a boom long enough to reach the top of a 30 story building. A truck crane (whether hydraulic or lattice boom) would not be well-suited for use on a high-rise project for the same reason.

Tower cranes are advantageous in high-rise construction due to their long booms, and due to the placement of the counterweight on the short "machinery" arm of the boom (rather than near the ground). While the lifting capacity of tower cranes is not as great as other types of cranes, it is sufficient for concrete, which is relatively light compared to other structural materials, such as steel.

The answer is (A).

Author Commentary

Information about cranes and their uses can be found from various sources, including *Cranes and Derricks* (see Codes and References).

14. Sumps and trenches are mainly used for controlling surface water, so they would not be suitable for this job. Wellpoints have a maximum lift of 25 ft, but usually have an effective lift of approximately 15 ft, and they dewater at a rate of approximately 100 gal/min, making them unsuitable for this job.

A deep well can be several hundred feet deep and can dewater at a rate of several thousand gallons per minute, making it the best dewatering method for this construction site.

The answer is (A).

Author Commentary

🕐 Eliminate options C and D first, since trenches and sumps are used only for controlling surface water.

15. The backhoe's productivity is

$$p_{\text{backhoe}} = E\left(\frac{c_h F I}{t}\right)$$

$$= \left(45 \ \frac{\text{min}}{\text{hr}}\right)\left(\frac{(1.5 \ \text{LCY})(0.80)(0.75)}{\frac{15 \ \text{sec}}{60 \ \frac{\text{sec}}{\text{min}}}}\right)$$

$$= 162 \ \text{BCY/hr} \quad (160 \ \text{BCY/hr})$$

The answer is (B).

Author Commentary

It is important to understand how to calculate the productivity of equipment, as it has a major influence on efficiency, and therefore profitability. Shorter cycle times equate to lower costs.

16. There are temporary solutions and long-term solutions to increasing productivity. Having laborers work overtime hours will temporarily increase daily output, but it is not a sustainable solution. Providing additional supervision is also likely to temporarily increase output, but since supervisors tend to move between jobsites as needed, it is not a sustainable solution, either. Of the options given, the most effective way of increasing the productivity of a crew is to provide more efficient equipment.

The answer is (B).

Author Commentary

Another effective way of increasing worker productivity is to provide workers with additional training.

17. According to Chap. 8 of the EPA's *Developing Your Stormwater Pollution Prevention Plan*, material stockpiles should never be placed on paved surfaces. Paved surfaces have a higher runoff coefficient than unimproved areas, and are less porous than gravel, dirt, or grassy areas. Placing a material stockpile on a paved surface can result in a fine or other disciplinary action.

The answer is (A).

18. Determine the project length with no delays, and identify the critical path (see *Illustration for Solution 18(a)*).

Determine the project duration accounting for delays in activities C, E, and G, and adjust the critical path (see *Illustration for Solution 18(b)*).

The answer is (D).

Author Commentary

Project sequencing is important in determining the critical path of the project. Once the plan and schedule are created, a change or delay in any activity (on or off the critical path) may impact the total project duration, as well as which activities are on the critical path.

Illustration for Solution 18(a)

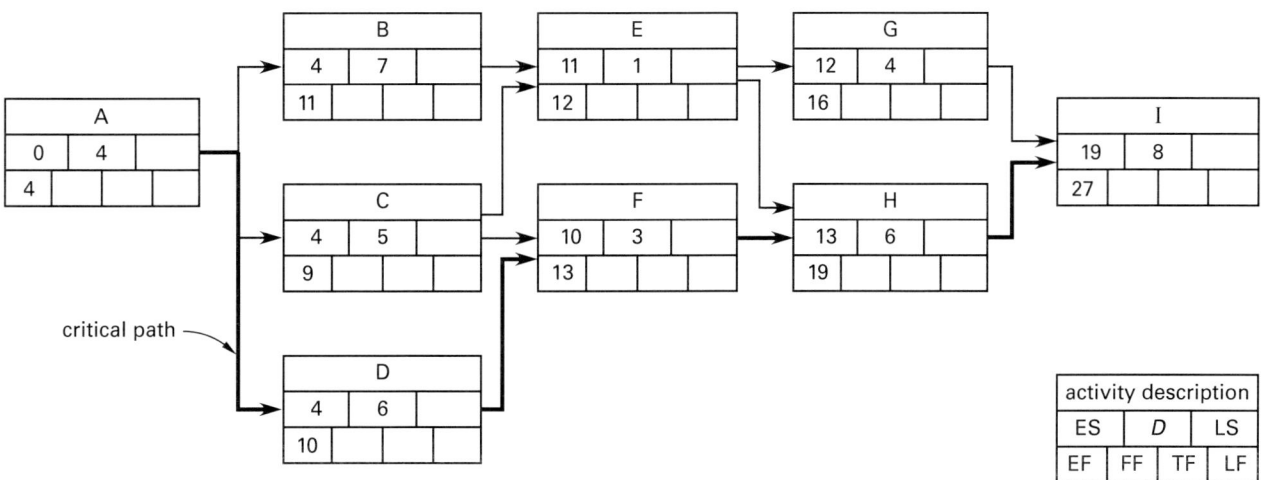

Illustration for Solution 18(b)

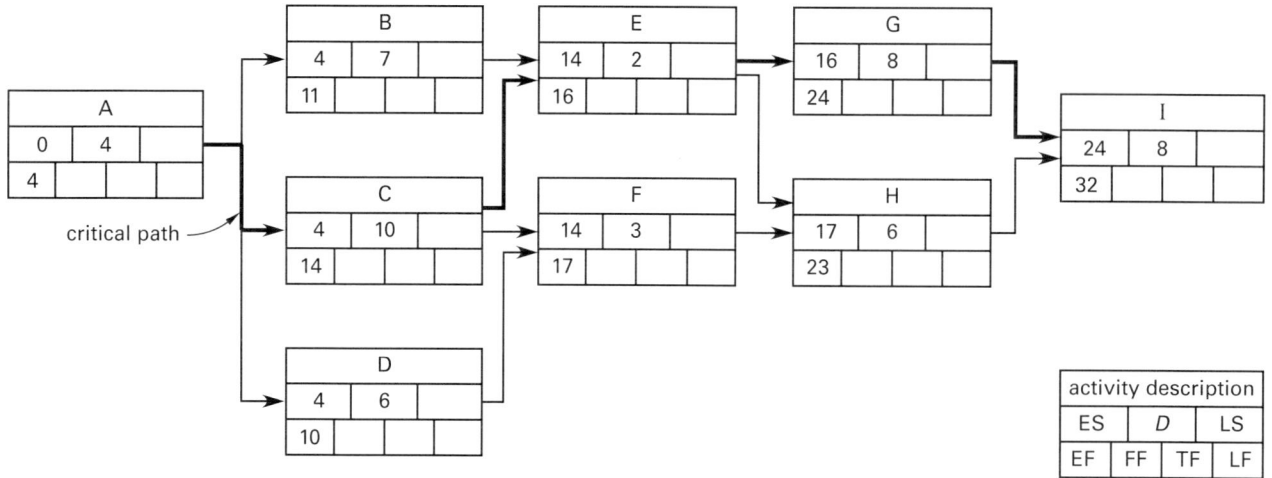

19. Create a critical path method (CPM) precedence diagram (see *Illustration for Solution 19*), beginning the project on day 0.

Complete a forward pass and determine the project duration.

For activity A, the earliest start (ES) value is 0 days because it is the first activity.

The earliest finish (EF) for activity A is

$$EF_A = ES_A + D_A$$
$$= 0 \text{ days} + 6 \text{ days}$$
$$= 6 \text{ days}$$

The ES value of activity B is 6 days, because the EF value of its only predecessor, activity A, is 6 days. The earliest finish for activity B is

$$EF_B = ES_B + D_B$$
$$= 6 \text{ days} + 8 \text{ days}$$
$$= 14 \text{ days}$$

Repeat these steps for each subsequent activity. Activities C, D, E, and F each have only one predecessor, making their earliest start values easy to determine.

For activity C,

$$ES_C = EF_A = 6 \text{ days}$$
$$EF_C = ES_C + D_C$$
$$= 6 \text{ days} + 7 \text{ days}$$
$$= 13 \text{ days}$$

For activity D,

$$ES_D = EF_B = 14 \text{ days}$$
$$EF_D = ES_D + D_D$$
$$= 14 \text{ days} + 3 \text{ days}$$
$$= 17 \text{ days}$$

For activity E,

$$ES_E = EF_C = 13 \text{ days}$$
$$EF_E = ES_E + D_E$$
$$= 13 \text{ days} + 6 \text{ days}$$
$$= 19 \text{ days}$$

For activity F,

$$ES_F = EF_D = 17 \text{ days}$$
$$EF_F = ES_F + D_F$$
$$= 17 \text{ days} + 3 \text{ days}$$
$$= 20 \text{ days}$$

When two or more activities precede a given activity, the largest EF value should be used to determine the ES of the activity. This can be seen with activities G, H, and I.

Activity G has two preceding activities, D and E. The EF of D is 17 days, and the EF of E is 19 days, so the ES of G is 19 days.

$$EF_G = ES_G + D_G$$
$$= 19 \text{ days} + 10 \text{ days}$$
$$= 29 \text{ days}$$

Illustration for Solution 19

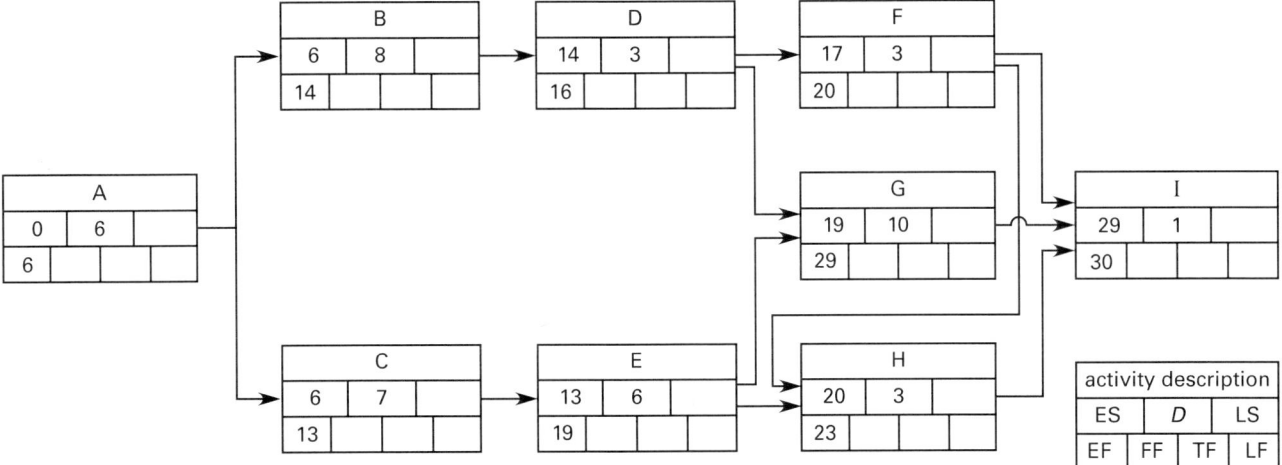

Activity H has two preceding activities, E and F. The EF of E is 19 days, and the EF of F is 20 days, so the ES of H is 20 days.

$$EF_H = ES_F + D_H = 20 \text{ days} + 3 \text{ days}$$
$$= 23 \text{ days}$$

Activity I has three predecessors, activity F, activity G, and activity H. The EF of F is 20 days, the EF of G is 29 days, and the EF of H is 23 days. Therefore, the ES of I is 29 days.

$$EF_I = ES_I + D_I = 29 \text{ days} + 1 \text{ day}$$
$$= 30 \text{ days}$$

Since activity I is the last task in the project, the earliest finish of the project is 30 days.

The answer is (D).

Author Commentary

- Read the problem thoroughly, and only complete the necessary passes. Only a forward pass is required to solve this problem.
- Don't forget that when two or more activities precede an activity, the *largest* earliest finish (EF) of the preceding activities determines the earliest start (ES) of that activity.

20. Free float (FF) is the amount of time an activity may be delayed without delaying the earliest start (ES) of its successor activity or activities.

Organize the activities in sequence based on information given in the table. When solving a critical path method (CPM) precedence diagram, the project begins on day 0. Perform a forward pass to determine the project duration (see *Illustration for Solution 20(a)*).

For activity A, the ES value is 0 days because it is the first activity. The earliest finish (EF) for activity A is

$$EF_A = ES_A + D_A = 0 \text{ days} + 5 \text{ days}$$
$$= 5 \text{ days}$$

The ES value of activity B is 5 days, because the EF value of its only predecessor, activity A, is 5 days. The EF for activity B is

$$EF_B = ES_B + D_B = 5 \text{ days} + 2 \text{ days}$$
$$= 7 \text{ days}$$

Activities C, D, E, and G each have only one predecessor, making their ES values easy to determine.

For activity C,

$$ES_C = EF_A = 5 \text{ days}$$
$$EF_C = ES_C + D_C$$
$$= 5 \text{ days} + 7 \text{ days}$$
$$= 12 \text{ days}$$

For activity D,

$$ES_D = EF_B = 5 \text{ days}$$
$$EF_D = ES_D + D_D$$
$$= 5 \text{ days} + 4 \text{ days}$$
$$= 9 \text{ days}$$

For activity E,

$$ES_E = EF_B = 7 \text{ days}$$
$$EF_E = ES_E + D_E$$
$$= 7 \text{ days} + 3 \text{ days}$$
$$= 10 \text{ days}$$

Illustration for Solution 20(a)

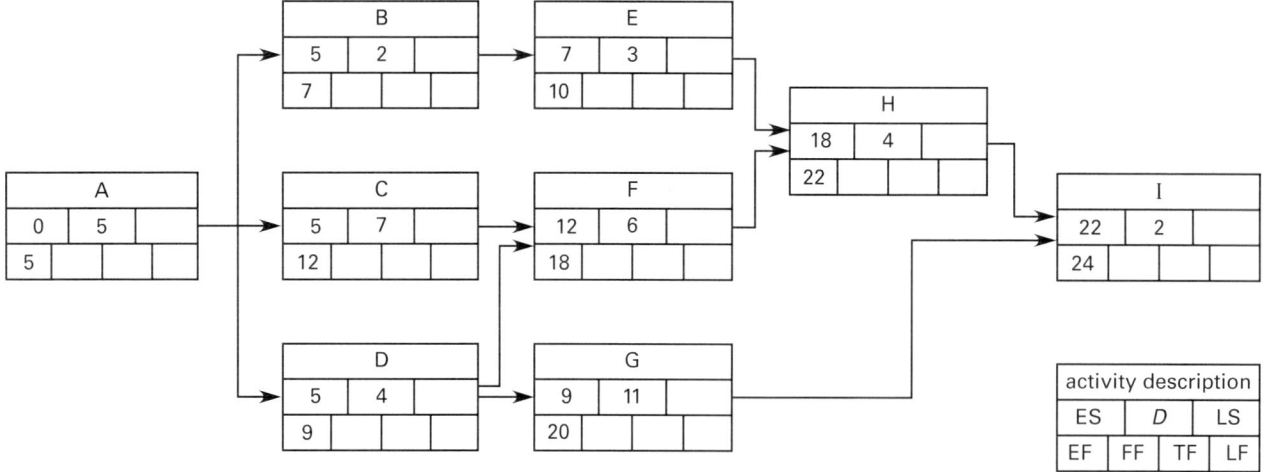

For activity G,

$$ES_G = EF_D = 9 \text{ days}$$
$$EF_G = ES_G + D_G$$
$$= 9 \text{ days} + 11 \text{ days}$$
$$= 20 \text{ days}$$

When two or more activities precede a given activity, the largest EF value should be used to determine the ES of the activity.

Activity F has two preceding activities, activity C and activity D. Activity C has an EF of 12 days, and activity D has an EF of 9 days, so the ES of activity F is 12 days.

$$EF_F = ES_F + D_F$$
$$= 12 \text{ days} + 6 \text{ days}$$
$$= 18 \text{ days}$$

Activity H also has two preceding activities, activity E and activity F. Activity E has an EF of 10 days, and activity F has an EF of 18 days, so the ES of activity H is 18 days.

$$EF_H = ES_H + D_H$$
$$= 18 \text{ days} + 4 \text{ days}$$
$$= 22 \text{ days}$$

Activity I has two preceding activities, activity G and activity H. Activity G has an EF of 20 days, and activity H has an EF of 22 days, so the ES of activity I is 22 days.

$$EF_I = ES_I + D_I$$
$$= 22 \text{ days} + 2 \text{ days}$$
$$= 24 \text{ days}$$

Complete a backward pass to determine the latest finish (LF) and latest start (LS) of each activity.

The LF of the last activity is equal to the EF of that activity. Since I is the last activity, its LF is equal to its EF, which is 24 days.

The LS of activity I is

$$LS_I = LF_I - D_I$$
$$= 24 \text{ days} - 2 \text{ days}$$
$$= 22 \text{ days}$$

Activities G, E, D, C, and B each have only one successor, making it simple to determine their LF and LS values.

Use the completed CPM diagram (see *Illustration for Solution 20(b)*) to determine the FF of each activity.

If an activity has multiple successors, use the smallest ES value of the successor activities.

For activity A,

$$FF_1 = ES_2 - EF_1$$
$$FF_A = ES_{B,C,D} - EF_A$$
$$= 5 \text{ days} - 5 \text{ days}$$
$$= 0$$

For activity B,

$$FF_B = ES_E - EF_B$$
$$= 7 \text{ days} - 7 \text{ days}$$
$$= 0$$

For activity C,

$$FF_C = ES_F - EF_C$$
$$= 12 \text{ days} - 12 \text{ days}$$
$$= 0$$

Illustration for Solution 20(b)

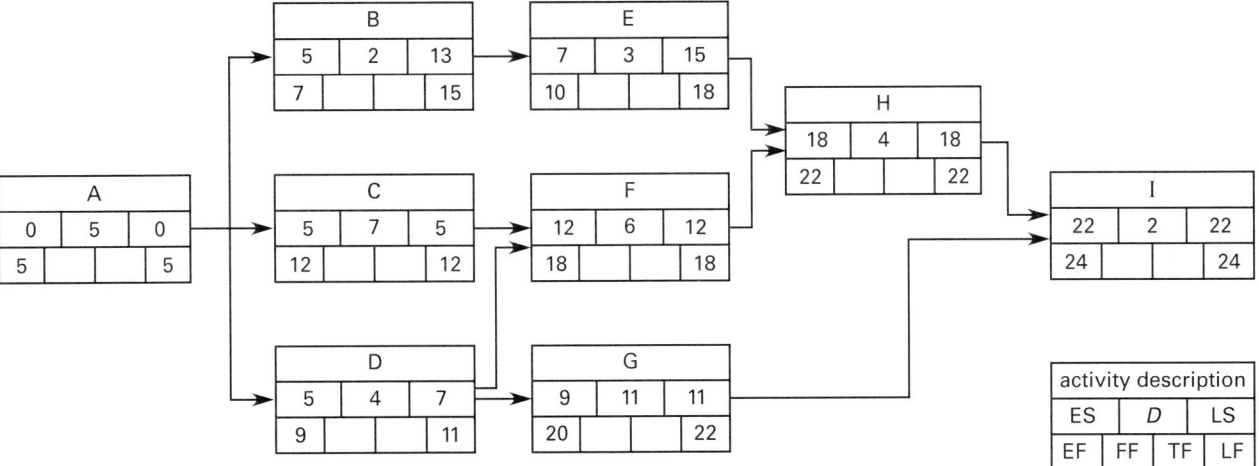

For activity D,
$$FF_D = ES_G - EF_D$$
$$= 9 \text{ days} - 9 \text{ days}$$
$$= 0$$

For activity E,
$$FF_E = ES_H - EF_E$$
$$= 18 \text{ days} - 10 \text{ days}$$
$$= 8 \text{ days}$$

For activity F,
$$FF_F = ES_H - EF_F$$
$$= 18 \text{ days} - 18 \text{ days}$$
$$= 0$$

For activity G,
$$FF_G = ES_I - EF_G$$
$$= 22 \text{ days} - 20 \text{ days}$$
$$= 2 \text{ days}$$

For activity H,
$$FF_H = ES_I - EF_H$$
$$= 22 \text{ days} - 22 \text{ days}$$
$$= 0$$

The FF of activity I is 0 days. The activity with the largest amount of free float is activity E with 8 days.

The answer is (C).

Author Commentary

💣 When completing a CPM diagram problem, remember that in a forward pass, the largest number rules, and in a backward pass, the smallest number rules.

21. Determine the total area to be primed and painted.

$$A_t = (5 \text{ rooms})(10{,}000 \text{ ft}^2) = 50{,}000 \text{ ft}^2$$

Calculate the number of working days required to apply the primer.

$$N_{\text{days,primer}} = \frac{50{,}000 \text{ ft}^2}{\left(1500 \dfrac{\text{ft}^2}{\text{hr}}\right)\left(8 \dfrac{\text{hr}}{\text{day}}\right)} = 4.167 \text{ days}$$

Calculate the number of working days required to apply the paint.

$$N_{\text{days,paint}} = \frac{(50{,}000 \text{ ft}^2)(2 \text{ coats})}{\left(1000 \dfrac{\text{ft}^2}{\text{hr}}\right)\left(8 \dfrac{\text{hr}}{\text{day}}\right)} = 12.5 \text{ days}$$

The total number of working days required to complete the priming and painting is

$$N_{\text{days},t} = N_{\text{days,primer}} + N_{\text{days,paint}}$$
$$= 4.167 \text{ days} + 12.5 \text{ days}$$
$$= 16.667 \text{ days} \quad (17 \text{ working days})$$

The answer is (C).

Author Commentary

💣 Be careful to note whether more than one coat of primer or paint is required.

22. Determine the actual cost of work performed, ACWP, for completed activities and partially completed activities.

$$ACWP = AC_{A,B,C} + AC_D + AC_E$$
$$= \$9500 + \$9000 + \$8500$$
$$= \$27{,}000$$

Determine the budgeted cost of work performed, BCWP, taking into account the percentage of work completed for partially completed tasks.

$$BCWP = BC_A + BC_B + BC_C + (0.75)(BC_D)$$
$$+ (0.5)(BC_E)$$
$$= \$2000 + \$1000 + \$4000 + (0.75)(\$14{,}000)$$
$$+ (0.5)(\$20{,}000)$$
$$= \$27{,}500$$

Determine the cost variance, CV, of the project.

$$CV = BCWP - ACWP$$
$$= \$27{,}500 - \$27{,}000$$
$$= \$500 \quad \text{[under budget]}$$

The answer is (A).

Author Commentary

Determining the cost variance, CV, of a project is an important tool used to make sure a project is on track financially. Cost variance is often associated with schedule variance, which uses a similar set of equations.

The equations for actual cost of work performed, budgeted cost of work performed, and cost variance may be found in the *Construction Depth Reference Manual*.

- A positive CV represents a project that is under budget. A negative CV represents a project that is over budget.

23. Resource leveling is a scheduling tool that helps managers balance the number of crews, amount of equipment, and amount of stored materials on a jobsite. If the schedule involves a great increase in the amount of resources needed followed by a decrease in resources needed, it will be difficult for crews and equipment to be used efficiently, so project managers try to distribute crews and materials equally throughout the schedule.

One assumption of resource leveling is that the project completion date cannot be changed. Waiting to start an activity that is on the critical path violates this assumption and, therefore, is not an example of resource leveling.

The answer is (C).

24. Determine how much the activity will cost if the crew does not work overtime.

$$C_{reg} = \left(1200 \ \frac{\$}{\text{day}}\right)\left(5 \ \frac{\text{day}}{\text{wk}}\right)(4 \ \text{wk})$$
$$= \$24{,}000$$

Determine how much the activity will cost if the crew does work overtime.

$$C_{OT} = \left(1200 \ \frac{\$}{\text{day}}\right)\left(5 \ \frac{\text{day}}{\text{wk}}\right)(3 \ \text{wk})$$
$$+ (1.5)\left(\frac{1200 \ \frac{\$}{\text{day}}}{8 \ \frac{\text{hr}}{\text{day}}}\right)\left(4 \ \frac{\text{OT hr}}{\text{day}}\right)$$
$$\times \left(5 \ \frac{\text{day}}{\text{wk}}\right)(3 \ \text{wk})$$
$$= \$31{,}500$$

Find the difference in cost.

$$\Delta C = C_{OT} - C_{reg} = \$31{,}500 - \$24{,}000$$
$$= \$7500$$

The answer is (B).

Author Commentary

Working overtime hours will often help a project get back on schedule, but in most cases it is extremely expensive. However, when faced with liquidated damages, it may be the best option. Make sure that calculations for overtime are completed using overtime rates to ensure accuracy of the analysis.

25. Consult a reference book such as *Foundation Design, Principles, and Practices* (see Codes and References) to find SPT hammer efficiency factors.

The corrected SPT *N*-value is

$$N_{60} = \frac{E_m C_B C_S C_R N}{0.6}$$
$$= \frac{(0.60)(1.15)(1.20)(0.95)(28)}{0.6}$$
$$= 36.7 \quad (37)$$

The answer is (C).

Author Commentary

- Be sure to correctly read all relevant tables. It is also important to remember to account for all of the variables when calculating the corrected *N*-value.

26. Welding symbols are used on drawings to show where welds are to be placed and other details, such as size, type, number of welds, and spacing. The symbols used in welding are shown.

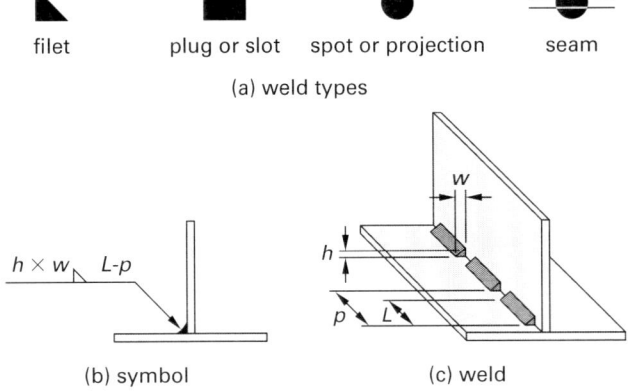

(a) weld types

(b) symbol   (c) weld

The answer is (D).

Author Commentary

The pitch of the weld, p, may be measured from the beginning of one weld to the beginning of the next weld, as shown in the illustration. The same measurement can be found by measuring from the centerline of one weld to the centerline of the next weld.

🕒 Tab your reference material to quickly find common weld symbols and charts.

27. Refer to Sec. 1205.2 of the IBC.

> **1205.2 Natural light.** The minimum net glazed area shall not be less than 8 percent of the floor of the room served.

Determine the total floor area of the building.

$$A_f = (100 \text{ ft})(30 \text{ ft}) + (50 \text{ ft})(20 \text{ ft}) = 4000 \text{ ft}^2$$

Determine the area of windows required.

$$A_{w,\text{req}} = (4000 \text{ ft}^2)(0.08) = 320 \text{ ft}^2$$

The area of one window is

$$A_w = (3 \text{ ft})(4 \text{ ft}) = 12 \text{ ft}^2$$

Therefore, the total number of windows required is

$$N_{w,t} = \frac{A_{w,\text{req}}}{A_w} = \frac{320 \text{ ft}^2}{12 \frac{\text{ft}^2}{\text{window}}}$$

$$= 26.67 \text{ windows} \quad (27 \text{ windows})$$

The answer is (C).

Author Commentary

When solving problems related to building design, the IBC is the most logical reference to consult.

🕒 Bring a copy of the IBC to the exam and tab where appropriate.

28. The concrete yield is the volume of freshly mixed concrete from a known quantity of ingredients. Calculate the volume of each ingredient based on saturated, surface-dry (SSD) density.

For cement,

$$V_{\text{absolute},c} = \frac{W}{(\text{SG})\gamma_w}$$

$$= \frac{(30 \text{ sacks})\left(94 \frac{\text{lbf}}{\text{sack}}\right)}{(3.15)\left(62.4 \frac{\text{lbf}}{\text{ft}^3}\right)}$$

$$= 14.35 \text{ ft}^3$$

For fine aggregate,

$$V_{\text{absolute,fa}} = \frac{W}{(\text{SG})\gamma_w}$$

$$= \frac{6000 \text{ lbf}}{(2.65)\left(62.4 \frac{\text{lbf}}{\text{ft}^3}\right)}$$

$$= 36.28 \text{ ft}^3$$

The correction for water deficit for fine aggregate is

$$C_{\text{fa}} = (-0.015)(6000 \text{ lbf}) = -90.0 \text{ lbf}$$

For coarse aggregate,

$$V_{\text{absolute,ca}} = \frac{W}{(\text{SG})\gamma_w}$$

$$= \frac{9500 \text{ lbf}}{(2.74)\left(62.4 \frac{\text{lbf}}{\text{ft}^3}\right)}$$

$$= 55.56 \text{ ft}^3$$

The correction for water excess for coarse aggregate is

$$C_{\text{ca}} = (0.02)(9500 \text{ lbf}) = 190.0 \text{ lbf}$$

For water,

$$V_{\text{absolute},w} = \frac{135 \text{ gal}}{7.48 \frac{\text{gal}}{\text{ft}^3}} = 18.0 \text{ ft}^3$$

The aggregate deviation from SSD correction is

$$\frac{C_{\text{fa}} + C_{\text{ca}}}{\gamma_w} = \frac{-90.0 \text{ lbf} + 190.0 \text{ lbf}}{62.4 \frac{\text{lbf}}{\text{ft}^3}} = 1.6 \text{ ft}^3$$

The volume of water is

$$V_w = 18.0 \text{ ft}^3 + 1.6 \text{ ft}^3 = 19.6 \text{ ft}^3$$

The volumetric concrete yield is

$$\frac{14.35 \text{ ft}^3 + 36.28 \text{ ft}^3 + 55.56 \text{ ft}^3 + 19.6 \text{ ft}^3}{(1-0.03)\left(3 \frac{\text{ft}}{\text{yd}}\right)^3}$$
$$= 4.80 \text{ yd}^3 \quad (4.8 \text{ yd}^3)$$

Alternatively, the volumetric concrete yield may be calculated by adding the entrained air percentage to one and multiplying this value by the summed volumes of the ingredients.

$$\frac{\left(\begin{array}{c}14.35 \text{ ft}^3 + 36.28 \text{ ft}^3 \\ + 55.56 \text{ ft}^3 + 19.6 \text{ ft}^3\end{array}\right)(1+0.03)}{\left(3 \frac{\text{ft}}{\text{yd}}\right)^3}$$
$$= 4.79 \text{ yd}^3 \quad (4.8 \text{ yd}^3)$$

The answer is (B).

Author Commentary

💣 Remember to take into account the water deficit or excess, and to convert from ft³ to yd³.

29. Calculate LRFD load combinations using ASCE 7 Sec. 2.3.

For load combination 1,

$$U = 1.4D = (1.4)\left(100 \frac{\text{lbf}}{\text{ft}^2}\right)$$
$$= 140 \text{ lbf/ft}^2$$

For load combination 2,

$$U = 1.2D + 1.6L + 0.5(L_r \text{ or } S \text{ or } R)$$
$$= (1.2)\left(100 \frac{\text{lbf}}{\text{ft}^2}\right) + (1.6)\left(50 \frac{\text{lbf}}{\text{ft}^2}\right) + (0.5)\left(20 \frac{\text{lbf}}{\text{ft}^2}\right)$$
$$= 210 \text{ lbf/ft}^2$$

For load combination 3,

$$U = 1.2D + 1.6(L_r \text{ or } S \text{ or } R) + (L \text{ or } 0.5W)$$
$$= (1.2)\left(100 \frac{\text{lbf}}{\text{ft}^2}\right) + (1.6)\left(20 \frac{\text{lbf}}{\text{ft}^2}\right) + 50 \frac{\text{lbf}}{\text{ft}^2}$$
$$= 202 \text{ lbf/ft}^2$$

For load combination 4,

$$U = 1.2D + 1.0W + L + 0.5(L_r \text{ or } S \text{ or } R)$$
$$= (1.2)\left(100 \frac{\text{lbf}}{\text{ft}^2}\right) + (1.0)\left(10 \frac{\text{lbf}}{\text{ft}^2}\right)$$
$$+ \left(50 \frac{\text{lbf}}{\text{ft}^2}\right) + (0.5)\left(20 \frac{\text{lbf}}{\text{ft}^2}\right)$$
$$= 190 \text{ lbf/ft}^2$$

For load combination 5,

$$U = 1.2D + 1.0E + L + 0.2S$$
$$= (1.2)\left(100 \frac{\text{lbf}}{\text{ft}^2}\right) + (1.0)\left(0 \frac{\text{lbf}}{\text{ft}^2}\right)$$
$$+ 50 \frac{\text{lbf}}{\text{ft}^2} + (0.2)\left(20 \frac{\text{lbf}}{\text{ft}^2}\right)$$
$$= 174 \text{ lbf/ft}^2$$

For load combination 6,

$$U = 0.9D + 1.0W$$
$$= (0.9)\left(100 \frac{\text{lbf}}{\text{ft}^2}\right) + (1.0)\left(10 \frac{\text{lbf}}{\text{ft}^2}\right)$$
$$= 100 \text{ lbf/ft}^2$$

For load combination 7,

$$U = 0.9D + 1.0E$$
$$= (0.9)\left(100 \frac{\text{lbf}}{\text{ft}^2}\right) + (1.0)\left(0 \frac{\text{lbf}}{\text{ft}^2}\right)$$
$$= 90 \text{ lbf/ft}^2$$

The maximum LRFD load is 210 lbf/ft² from load combination 2. Calculate ASD load combinations using ASCE 7 Sec. 2.4.

For load combination 1,

$$U = D$$
$$= 100 \text{ lbf/ft}^2$$

For load combination 2,

$$U = D + L$$
$$= 100 \frac{\text{lbf}}{\text{ft}^2} + 50 \frac{\text{lbf}}{\text{ft}^2}$$
$$= 150 \text{ lbf/ft}^2$$

For load combination 3,

$$U = D + (L_r \text{ or } S \text{ or } R)$$
$$= 100 \frac{\text{lbf}}{\text{ft}^2} + 20 \frac{\text{lbf}}{\text{ft}^2}$$
$$= 120 \text{ lbf/ft}^2$$

For load combination 4,

$$U = D + 0.75L + 0.75(L_r \text{ or } S \text{ or } R)$$
$$= 100 \ \frac{\text{lbf}}{\text{ft}^2} + (0.75)\left(50 \ \frac{\text{lbf}}{\text{ft}^2}\right) + (0.75)\left(20 \ \frac{\text{lbf}}{\text{ft}^2}\right)$$
$$= 152.5 \ \text{lbf/ft}^2$$

For load combination 5,

$$U = D + (0.6W \text{ or } 0.7E)$$
$$= 100 \ \frac{\text{lbf}}{\text{ft}^2} + (0.6)\left(10 \ \frac{\text{lbf}}{\text{ft}^2}\right)$$
$$= 106 \ \text{lbf/ft}^2$$

For load combination 6a,

$$U = D + 0.75L + 0.75(0.6W)$$
$$\quad + 0.75(L_r \text{ or } S \text{ or } R)$$
$$= 100 \ \frac{\text{lbf}}{\text{ft}^2} + (0.75)\left(50 \ \frac{\text{lbf}}{\text{ft}^2}\right)$$
$$\quad + (0.75)(0.6)\left(10 \ \frac{\text{lbf}}{\text{ft}^2}\right) + (0.75)\left(20 \ \frac{\text{lbf}}{\text{ft}^2}\right)$$
$$= 157 \ \text{lbf/ft}^2$$

For load combination 6b,

$$U = D + 0.75L + 0.75(0.7E) + 0.75S$$
$$= 100 \ \frac{\text{lbf}}{\text{ft}^2} + (0.75)\left(50 \ \frac{\text{lbf}}{\text{ft}^2}\right)$$
$$\quad + (0.75)(0.7)\left(0 \ \frac{\text{lbf}}{\text{ft}^2}\right) + (0.75)\left(20 \ \frac{\text{lbf}}{\text{ft}^2}\right)$$
$$= 152.5 \ \text{lbf/ft}^2$$

For load combination 7,

$$U = 0.6D + 0.6W$$
$$= (0.6)\left(100 \ \frac{\text{lbf}}{\text{ft}^2}\right) + (0.6)\left(10 \ \frac{\text{lbf}}{\text{ft}^2}\right)$$
$$= 66 \ \text{lbf/ft}^2$$

For load combination 8,

$$U = 0.6D + 0.7E$$
$$= (0.6)\left(100 \ \frac{\text{lbf}}{\text{ft}^2}\right) + (0.7)\left(0 \ \frac{\text{lbf}}{\text{ft}^2}\right)$$
$$= 60 \ \text{lbf/ft}^2$$

The maximum ASD load is 157 lbf/ft² from load combination 6a. The difference between the maximum values calculated using LRFD and ASD, respectively, is

$$210 \ \frac{\text{lbf}}{\text{ft}^2} - 157 \ \frac{\text{lbf}}{\text{ft}^2} = 53 \ \text{lbf/ft}^2$$

The answer is (A).

Author Commentary

- LRFD and ASD factored loads are available in many different references, including the AISC *Steel Construction Manual* and ASCE 7.
- When there is an "or" in the factored load calculations, the larger value must be used.

30. According to OSHA Sec. 1926.451(a)(1), scaffolding should be designed to support, without failure, a minimum of its own weight and four times its maximum intended load.

Determine the maximum factored load.

$$w_u = w_{\max}(\text{FS}) = (500 \ \text{lbf})(4)$$
$$= 2000 \ \text{lbf}$$

Sum the factored load and the self-weight of the scaffolding.

$$2000 \ \text{lbf} + 200 \ \text{lbf} = 2200 \ \text{lbf}$$

The answer is (C).

Author Commentary

- Only the applied load is multiplied by the factor of safety. Multiplying both the applied load and self-weight by the factor of safety will result in an overly conservative answer.

31. According to ASTM C1074, *Standard Practice for Estimating Concrete Strength by the Maturity Method*, Sec. 5.3, options B, C, and D are limitations of the maturity method. Per ASTM C1074 Sec. 1.2, the concrete being tested is not required to be maintained at the same conditions as the concrete placed in the field.

The answer is (A).

Author Commentary

Typical concrete strength tests are completed on samples that maintain approximately the same conditions as the concrete for which strength is to be estimated. The maturity method allows laboratory-cured specimens to be tested and their strength to be extrapolated to the strength of concrete in the field. This is achieved through calculations given in ASTM C1074.

32. Use ASCE 7 Eq. 27.3-1, neglecting the values for K_z, K_{zt}, and K_d. The constant, 0.00256, is the mass density of air for the standard atmosphere and sea level pressure, and serves to convert wind speed to a load.

From ASCE 7 Eq. 27.3-1,

$$q_z = 0.00256 K_z K_{zt} K_d v^2 = 0.00256(v)^2$$
$$= (0.00256)\left(90 \; \frac{\text{mi}}{\text{hr}}\right)^2$$
$$= 20.74 \; \text{lbf/ft}^2$$

Find the wind load.

$$W = \left(20.74 \; \frac{\text{lbf}}{\text{ft}^2}\right)(6 \; \text{ft}) = 124.44 \; \text{lbf/ft}$$

Analyze the formwork as a simply supported member, and find the horizontal force on the top of the wall.

$$H_{\text{top}} = \frac{hW}{2} = \frac{(15 \; \text{ft})\left(124.44 \; \frac{\text{lbf}}{\text{ft}}\right)}{2}$$
$$= 933.3 \; \text{lbf} \quad (935 \; \text{lbf})$$

The answer is (C).

Author Commentary

- Be sure to convert values to the units the problem requires.
- The horizontal force is distributed evenly between the top and the bottom of the wall.

33. Evening the rock surface before the sheet piles are set and placing sacks of concrete or a thick layer of clay on the outside of the cofferdam cells are effective methods of preventing underflow from a sheet pile cofferdam on a rock.

Increasing the size of the dewatering machinery to counteract the increase in underflow is neither practical nor economical. It is costly, and underflow can result from openings under sheet piles.

The answer is (A).

Author Commentary

There are several other methods that may be used to control underflow. The bottom end of the pile, or toe, may be driven or bored into various layers of soil and/or rock. This process is called "toeing." Toeing the sheet piles into a graded layer of low permeability sand and clay can control, but not eliminate, underflow. Using cast-steel protectors on the tips of the sheet piles to permit toeing into rock can also help to prevent underflow.

34. According to OSHA Sec. 1926.452(c)(6), scaffolds 125 ft or higher above their base plates must be designed by registered professional engineers.

The answer is (D).

Author Commentary

- Reviewing OSHA Part 1926 in detail before the exam and adding tabs for quick reference will make finding relevant information during the exam easier.

35. Determine how far the base of the ladder must be from the building. OSHA Sec. 1926.1053(b)(5)(i) states that the horizontal distance from the top support to the base of the ladder must be approximately $1/4$ the working length.

$$D = \tfrac{1}{4}h = \left(\tfrac{1}{4}\right)(18 \; \text{ft})$$
$$= 4.5 \; \text{ft}$$

Calculate the hypotenuse of the triangle made by the horizontal distance and the height of the building.

$$L = \sqrt{h^2 + D^2} = \sqrt{(18 \; \text{ft})^2 + (4.5 \; \text{ft})^2}$$
$$= 18.6 \; \text{ft}$$

OSHA Sec. 1926.1053(b)(1) states that ladders used to access roofs must extend at least 3 ft above the roofline. Add this to the ladder length.

$$L = 18.6 \; \text{ft} + 3 \; \text{ft}$$
$$= 21.6 \; \text{ft} \quad (22 \; \text{ft})$$

The answer is (B).

Author Commentary

OSHA rules and regulations are intended to protect workers, so they are strictly enforced. Failure to adhere to OSHA regulations can result in large fines.

- Reviewing OSHA Part 1926 in detail before the exam and adding tabs for quick reference will make finding relevant information during the exam easier.

36. Determine the total number of hours the employees worked.

$$N_{\text{hr},t} = \left(50 \; \frac{\text{hr}}{\text{wk}}\right)(40 \; \text{wk})(500 \; \text{employees})$$
$$= 1{,}000{,}000 \; \text{hr}$$

The company's incidence rate is

$$\text{incidence rate} = \frac{\left(\begin{array}{c}\text{total no. of injuries}\\\text{and illnesses}\end{array}\right) \times (200{,}000 \; \text{hr})}{\text{hours worked by all employees}}$$
$$= \frac{(10)(200{,}000 \; \text{hr})}{1{,}000{,}000 \; \text{hr}}$$
$$= 2$$

The answer is (B).

Author Commentary

The 200,000 hours in the incidence rate equation represents the equivalent of 100 employees working 40 hours per week, 50 weeks per year, and provides the standard base for incidence rates.

37. Find the peak flow rate using the runoff coefficients as the velocity of flow.

$$Q = A_t v$$
$$= A_1 c_1 + A_2 c_2 + A_3 c_3 + A_4 c_4$$
$$= \frac{\begin{pmatrix}(4 \text{ ac})(0.20) + (6 \text{ ac})(0.10) \\ + (4 \text{ ac})(0.25) + (10 \text{ ac})(0.35)\end{pmatrix} \times \left(43{,}560 \frac{\text{ft}^2}{\text{ac}}\right)\left(7.48 \frac{\text{gal}}{\text{ft}^3}\right)\left(3.5 \frac{\text{in}}{\text{hr}}\right)}{\left(60 \frac{\text{min}}{\text{hr}}\right)\left(12 \frac{\text{in}}{\text{ft}}\right)}$$
$$= 9345 \text{ gal/min}$$

Calculate the required pump power.

$$P_p = \frac{Q h_{td}}{3956 \frac{\text{ft-gal}}{\text{hp-min}}} = \frac{\left(9345 \frac{\text{gal}}{\text{min}}\right)(8 \text{ ft})}{3956 \frac{\text{ft-gal}}{\text{hp-min}}}$$
$$= 18.90 \text{ hp}$$

The required brake horsepower is

$$\text{BHP} = \frac{P_p}{\eta} = \frac{18.90 \text{ hp}}{0.85}$$
$$= 22.24 \text{ hp} \quad (22 \text{ hp})$$

The answer is (D).

Author Commentary

- Consult the *Construction Depth Reference Manual* for information on and equations for construction dewatering and pumping.
- Remember to adjust for the efficiency of the pump.

38. Check the pile spacing. Piles with spacing of at least three times the pile diameter can be assumed to act individually, not as a block (i.e., the pile group efficiency is 100%). The actual spacing is 4.5 ft. The minimum pile spacing for individual action is

$$(3)\left(\frac{18 \text{ in}}{12 \frac{\text{in}}{\text{ft}}}\right) = 4.5 \text{ ft} \quad [\text{OK}]$$

For pile caps that do not bear on the surface, the ultimate static capacity consists of the end bearing capacity (negligible in this case) and the shaft's surface area friction capacity. The capacity is reduced by the weights of the pile cap and piles. Although the specific weight of plain concrete is approximately 140–145 lbf/ft³, the specific weight of reinforced concrete is approximately 150 lbf/ft³.

$$Q_{\text{design}} = \frac{Q_u}{\text{FS}} = \frac{Q_{\text{tip bearing}} + Q_{\text{shaft friction}} - W_{\text{cap}} - W_{\text{piles}}}{\text{FS}}$$
$$= \frac{0 + n A_{\text{barrel}} \sigma_f - V_{\text{cap}} \gamma_{\text{cap}} - V_{\text{piles}} \gamma_{\text{piles}}}{\text{FS}}$$
$$= \frac{n \pi d L \sigma_f - \left(B^2 t + n \frac{\pi}{4} d^2 L\right) \gamma_{\text{concrete}}}{\text{FS}}$$
$$= \frac{\dfrac{(4\pi)(18 \text{ in})(50 \text{ ft})\left(800 \frac{\text{lbf}}{\text{ft}^2}\right)}{12 \frac{\text{in}}{\text{ft}}} - \begin{pmatrix}(6 \text{ ft})^2(2 \text{ ft}) \\ + \dfrac{(4)\left(\frac{\pi}{4}\right)(18 \text{ in})^2(50 \text{ ft})}{\left(12 \frac{\text{in}}{\text{ft}}\right)^2}\end{pmatrix}\left(150 \frac{\text{lbf}}{\text{ft}^3}\right)}{(3)\left(2000 \frac{\text{lbf}}{\text{ton}}\right)}$$
$$= 115 \text{ tons} \quad (110 \text{ tons})$$

The answer is (C).

Author Commentary

In this solution, buoyancy has been disregarded. The effect of buoyancy on the supported weight is small. (It could also be argued that the self-weight of the piles should be disregarded, since the specific weight of concrete is almost equal to the specific weight of the soil displaced.) Since the end (tip) bearing capacity is considered negligible, Terzaghi's bearing capacity factors are not needed for the bearing capacity calculation. Since the soil density only appears in the end bearing capacity calculation (which is not used), the location of the water table only affects the pile weight. There are many factors that can complicate pile and pile group capacities. For example, the distribution of frictional forces on the pile would normally depend on soil properties, including a critical depth. The calculation of pile group capacities in cohesive soils is even more complex.

- Don't forget to subtract the weight of the pile cap.

39. The expected elongation for each rod is

$$\delta = \frac{L_o F}{EA}$$

$$= \frac{(5 \text{ ft})\left(12 \frac{\text{in}}{\text{ft}}\right)\left(\dfrac{1000 \text{ lbf} \left(\dfrac{1000 \frac{\text{lbf}}{\text{kip}}}{2}\right)}{}\right)}{\left(7500 \frac{\text{kips}}{\text{in}^2}\right)\pi\left(\dfrac{1 \text{ in}}{2}\right)^2}$$

$$= 0.005 \text{ in}$$

The answer is (B).

Author Commentary

- Make sure to use conversion factors when necessary.
- It is important to remember to divide the force by 2 since the weight of the sign is shared equally between the two rods.

40. Use ACI 347 Table 2.1 to determine the unit weight coefficient, C_w. For concrete with a unit weight of 145 lbf/ft^3, the unit weight coefficient is 1.0.

Use ACI 347 Table 2.2 to determine the chemistry coefficient, C_c. For type II concrete without admixtures (retarders) that increase hardening time, the chemistry coefficient is 1.0.

Use ACI 347 Eq. 2.4 to determine the maximum lateral pressure by the wall.

$$p_{\max} = C_w C_c \left(150 + \frac{43{,}400}{T} + 2800\left(\frac{R}{T}\right)\right)$$

$$= (1.0)(1.0)\left(150 + \frac{43{,}400}{75°\text{F}} + (2800)\left(\frac{10 \frac{\text{ft}}{\text{hr}}}{75°\text{F}}\right)\right)$$

$$= 1102 \text{ lbf/ft}^2 \quad (1100 \text{ lbf/ft}^2)$$

The answer is (B).

Author Commentary

- The maximum pressure affects how the formwork is designed. However, the formwork construction details are not relevant to the calculation of maximum pressure.
- The temperature of the concrete, not the air temperature, determines the speed of the hardening process.

Practice Exam 2 Answer Key

41. C	51. B	61. D	71. B
42. B	52. B	62. D	72. A
43. A	53. B	63. A	73. B
44. A	54. C	64. C	74. D
45. C	55. D	65. B	75. C
46. D	56. B	66. D	76. B
47. C	57. B	67. A	77. C
48. D	58. D	68. C	78. B
49. C	59. D	69. B	79. D
50. D	60. B	70. C	80. A

Solutions
Practice Exam 2

41. Calculate the total weight of soil needed on the jobsite.

$$W = V_s \gamma_s = (120{,}000 \text{ yd}^3)\left(135 \frac{\text{lbf}}{\text{ft}^3}\right)\left(3 \frac{\text{ft}}{\text{yd}}\right)^3$$
$$= 437{,}400{,}000 \text{ lbf}$$

Calculate the volume of soil needed from the borrow pit.

$$V_{\text{BCY}} = \frac{437{,}400{,}000 \text{ lbf}}{\left(120 \frac{\text{lbf}}{\text{ft}^3}\right)\left(3 \frac{\text{ft}}{\text{yd}}\right)^3}$$
$$= 135{,}000 \text{ yd}^3 \quad (135{,}000 \text{ BCY})$$

The answer is (C).

Alternative Solution

Find the ratio of the jobsite soil density to the borrow pit soil density.

$$\frac{\gamma_{s,\text{jobsite}}}{\gamma_{s,\text{borrow pit}}} = \frac{135 \frac{\text{lbf}}{\text{ft}^3}}{120 \frac{\text{lbf}}{\text{ft}^3}} = 1.125$$

Find the volume of soil needed from the borrow pit.

$$V_{\text{BCY}} = (120{,}000 \text{ yd}^3)(1.125)$$
$$= 135{,}000 \text{ yd}^3 \quad (135{,}000 \text{ BCY})$$

The answer is (C).

Author Commentary

Carefully read the problem and determine the fastest way to calculate the answer.

- ⏱ The unit weight ratio method may be used when the unit weight of one material and factor are given, and the unit weight of the second material needs to be calculated. If adequate information is provided, use this alternative method to save time.
- 💣 Be careful to correctly convert units into yd^3.

42. Use the weighted average method. Find the average elevation of the existing points.

$$\text{elev}_{\text{ave}} = \frac{\sum \text{elev}}{N_{\text{points}}} = \frac{\begin{array}{l}101.5 \text{ ft} + 102.0 \text{ ft} + 103.0 \text{ ft}\\ + 102.5 \text{ ft} + 100.0 \text{ ft} + 101.5 \text{ ft}\\ + 102.5 \text{ ft} + 101.5 \text{ ft} + 100.5 \text{ ft}\\ + 101.0 \text{ ft} + 101.5 \text{ ft} + 101.0 \text{ ft}\end{array}}{12 \text{ points}}$$
$$= 101.5 \text{ ft}$$

Subtract the required elevation from the average existing elevation to find the change in elevation.

$$\Delta_{\text{elev}} = 101.5 \text{ ft} - 95.0 \text{ ft} = 6.5 \text{ ft}$$

Find the total volume of soil to be excavated.

$$V_t = hLN_{\text{blocks}}D_{\text{ave}} = \frac{(50 \text{ ft})(50 \text{ ft})(6 \text{ blocks})(6.5 \text{ ft})}{\left(3 \frac{\text{ft}}{\text{yd}}\right)^3}$$
$$= 3611 \text{ yd}^3 \quad (3600 \text{ yd}^3)$$

The answer is (B).

Author Commentary

The average depth method is often used when information about topography is available. It is a more precise measure than other methods, but it often requires a larger data set in order to be accurate.

- ⏱ Using the weighted average of the elevation differences is the key to solving this type of problem more quickly.

43. Refer to OSHA Sec. 1926.752(c).

Site layout. The controlling contractor shall ensure that the following is provided and maintained:

1926.752(c)(1)

Adequate access roads into and through the site for the safe delivery and movement of derricks, cranes, trucks, other necessary equipment, and the material to be erected and means and methods for pedestrian and vehicular control. Exception: this requirement does not apply to roads outside of the construction site.

1926.752(c)(2)

A firm, properly graded, drained area, readily accessible to the work with adequate space for the safe storage of materials and the safe operation of the erector's equipment.

Basic construction jobsite layout principles stress controlling access. Visitors are required to stop at the jobsite trailer when entering a construction site.

The answer is (A).

44. Determine the volume of cut required for each station increment.

From sta 0 to sta 1,

$$V_{c,0-1} = L\left(\frac{A_{c,0} + A_{c,1}}{2}\right) = \frac{(100 \text{ ft})\left(\frac{45 \text{ ft}^2 + 50 \text{ ft}^2}{2}\right)}{\left(3 \frac{\text{ft}}{\text{yd}}\right)^3}$$

$$= 175.9 \text{ yd}^3$$

From sta 1 to sta 2,

$$V_{c,1-2} = L\left(\frac{A_{c,1} + A_{c,2}}{2}\right) = \frac{(100 \text{ ft})\left(\frac{50 \text{ ft}^2 + 57 \text{ ft}^2}{2}\right)}{\left(3 \frac{\text{ft}}{\text{yd}}\right)^3}$$

$$= 198.1 \text{ yd}^3$$

From sta 2 to sta 3,

$$V_{c,2-3} = L\left(\frac{A_{c,2} + A_{c,3}}{2}\right) = \frac{(100 \text{ ft})\left(\frac{57 \text{ ft}^2 + 23 \text{ ft}^2}{2}\right)}{\left(3 \frac{\text{ft}}{\text{yd}}\right)^3}$$

$$= 148.1 \text{ yd}^3$$

From sta 3 to sta 4,

$$V_{c,3-4} = L\left(\frac{A_{c,3} + A_{c,4}}{2}\right) = \frac{(100 \text{ ft})\left(\frac{23 \text{ ft}^2 + 0 \text{ ft}^2}{2}\right)}{\left(3 \frac{\text{ft}}{\text{yd}}\right)^3}$$

$$= 42.6 \text{ yd}^3$$

From sta 4 to sta 5,

$$V_{c,4-5} = L\left(\frac{A_{c,4} + A_{c,5}}{2}\right) = \frac{(100 \text{ ft})\left(\frac{0 \text{ ft}^2 + 0 \text{ ft}^2}{2}\right)}{\left(3 \frac{\text{ft}}{\text{yd}}\right)^3}$$

$$= 0$$

Determine the volume of fill for each station increment, taking shrinkage into account.

For sta 0 to sta 1,

$$V_{f,0-1} = L\left(\frac{A_{f,0} + A_{f,1}}{2}\right)(\text{shrinkage})$$

$$= \left(\frac{(100 \text{ ft})(20 \text{ ft}^2 + 13 \text{ ft}^2)}{(2)\left(3 \frac{\text{ft}}{\text{yd}}\right)^3}\right)(1.2)$$

$$= 73.3 \text{ yd}^3$$

For sta 1 to sta 2,

$$V_{f,1-2} = L\left(\frac{A_{f,1} + A_{f,2}}{2}\right)(\text{shrinkage})$$

$$= \left(\frac{(100 \text{ ft})(13 \text{ ft}^2 + 0 \text{ ft}^2)}{(2)\left(3 \frac{\text{ft}}{\text{yd}}\right)^3}\right)(1.2)$$

$$= 28.9 \text{ yd}^3$$

For sta 2 to sta 3,

$$V_{f,2-3} = L\left(\frac{A_{f,2} + A_{f,3}}{2}\right)(\text{shrinkage})$$

$$= \left(\frac{(100 \text{ ft})(0 \text{ ft}^2 + 30 \text{ ft}^2)}{(2)\left(3 \frac{\text{ft}}{\text{yd}}\right)^3}\right)(1.2)$$

$$= 66.7 \text{ yd}^3$$

For sta 3 to sta 4,

$$V_{f,3-4} = L\left(\frac{A_{f,3} + A_{f,4}}{2}\right)(\text{shrinkage})$$

$$= \left(\frac{(100 \text{ ft})(30 \text{ ft}^2 + 29 \text{ ft}^2)}{(2)\left(3 \dfrac{\text{ft}}{\text{yd}}\right)^3}\right)(1.2)$$

$$= 131.1 \text{ yd}^3$$

For sta 4 to sta 5,

$$V_{f,4-5} = L\left(\frac{A_{f,4} + A_{f,5}}{2}\right)(\text{shrinkage})$$

$$= \left(\frac{(100 \text{ ft})(29 \text{ ft}^2 + 23 \text{ ft}^2)}{(2)\left(3 \dfrac{\text{ft}}{\text{yd}}\right)^3}\right)(1.2)$$

$$= 115.6 \text{ yd}^3$$

Determine the total volume for each increment. Represent cut as a positive number and fill as a negative number.

For sta 0 to sta 1,

$$V_{t,0-1} = V_{c,0-1} - V_{f,0-1} = 175.9 \text{ yd}^3 - 73.3 \text{ yd}^3$$
$$= 102.6 \text{ yd}^3$$

For sta 1 to sta 2,

$$V_{t,1-2} = V_{c,1-2} - V_{f,1-2} = 198.1 \text{ yd}^3 - 28.9 \text{ yd}^3$$
$$= 169.2 \text{ yd}^3$$

From sta 2 to sta 3,

$$V_{t,2-3} = V_{c,2-3} - V_{f,2-3} = 148.1 \text{ yd}^3 - 66.7 \text{ yd}^3$$
$$= 81.4 \text{ yd}^3$$

From sta 3 to sta 4,

$$V_{t,3-4} = V_{c,3-4} - V_{f,3-4} = 42.6 \text{ yd}^3 - 131.1 \text{ yd}^3$$
$$= -88.5 \text{ yd}^3$$

From sta 4 to sta 5,

$$V_{t,4-5} = V_{c,4-5} - V_{f,4-5} = 0 \text{ yd}^3 - 115.6 \text{ yd}^3$$
$$= -115.6 \text{ yd}^3$$

Add all total volumes to determine the net cut or fill required for the project.

$$V_{\text{net}} = V_{t,0-1} + V_{t,1-2} + V_{t,2-3} + V_{t,3-4} + V_{t,4-5}$$
$$= 102.6 \text{ yd}^3 + 169.2 \text{ yd}^3 + 81.4 \text{ yd}^3$$
$$\quad - 88.5 \text{ yd}^3 - 115.6 \text{ yd}^3$$
$$= 149.1 \text{ yd}^3 \quad (150 \text{ yd}^3 \text{ cut})$$

150 yd^3 of cut are required.

The answer is (A).

Author Commentary

Earthwork end area mass diagrams are a useful tool in conceptual estimates. Often, construction work is performed on a unit price basis. Both cut and fill volumes are important because they relate directly to payment.

💣 A shrinkage factor is only used to adjust fill, not cut.

45. Determine how many square feet of brick are required for the north and west facing walls.

$$A_{b,N,W} = (200 \text{ ft} + 50 \text{ ft} + 150 \text{ ft})(4 \text{ ft}) = 1600 \text{ ft}^2$$

Determine how many square feet of brick are required for the south and east facing walls.

$$A_{b,S,E} = (250 \text{ ft} + 100 \text{ ft} + 50 \text{ ft})(16 \text{ ft}) = 6400 \text{ ft}^2$$

Calculate the area of one brick.

$$A_b = \frac{(8 \text{ in})(4 \text{ in})}{\left(12 \dfrac{\text{in}}{\text{ft}}\right)^2} = 0.22 \text{ ft}^2$$

Find the number of bricks required.

$$N_{b,\text{req}} = \frac{A_w}{A_b} = \frac{1600 \text{ ft}^2 + 6400 \text{ ft}^2}{0.22 \text{ ft}^2}$$
$$= 36{,}364 \quad (36{,}400 \text{ bricks})$$

The answer is (C).

Author Commentary

Calculating the number of bricks required for construction projects requires knowledge about brick sizes, types of bonds (bricklaying positions), and mortar width. In this problem, mortar width is included with brick size, thereby eliminating the need for additional calculations.

46. Determine the total weight of each type of channel, beam, and column. The weights of these members can be found in the AISC *Steel Construction Manual*.

For the channels,

$$W_{t,\text{MC12}} = Q_c L_c w_c = (8)(12 \text{ ft})\left(40 \ \frac{\text{lbf}}{\text{ft}}\right) = 3840 \text{ lbf}$$

$$W_{t,\text{MC8}} = Q_c L_c w_c = (6)(10 \text{ ft})\left(22.8 \ \frac{\text{lbf}}{\text{ft}}\right) = 1368 \text{ lbf}$$

For the beams,

$$W_{t,\text{W12}} = Q_b L_b w_b = (12)(12 \text{ ft})\left(26 \ \frac{\text{lbf}}{\text{ft}}\right) = 3744 \text{ lbf}$$

$$W_{t,\text{W14}} = Q_b L_b w_b = (4)(24 \text{ ft})\left(26 \ \frac{\text{lbf}}{\text{ft}}\right) = 2496 \text{ lbf}$$

$$W_{t,\text{W24}} = Q_b L_b w_b = (6)(24 \text{ ft})\left(84 \ \frac{\text{lbf}}{\text{ft}}\right) = 12{,}096 \text{ lbf}$$

$$W_{t,\text{W8}} = Q_b L_b w_b = (9)(10 \text{ ft})\left(28 \ \frac{\text{lbf}}{\text{ft}}\right) = 2520 \text{ lbf}$$

For the columns,

$$W_{t,\text{HSS8}} = Q_c L_c w_c = (14)(16 \text{ ft})\left(17.28 \ \frac{\text{lbf}}{\text{ft}}\right)$$
$$= 3871 \text{ lbf}$$

$$W_{t,\text{HSS14}} = Q_c L_c w_c = (4)(14 \text{ ft})\left(68.24 \ \frac{\text{lbf}}{\text{ft}}\right)$$
$$= 3821 \text{ lbf}$$

$$W_{t,\text{HSS16}} = Q_c L_c w_c = (2)(14 \text{ ft})\left(127.00 \ \frac{\text{lbf}}{\text{ft}}\right)$$
$$= 3556 \text{ lbf}$$

The total weight of steel needed is

$$W_t = W_{t,\text{MC12}} + W_{t,\text{MC8}} + W_{t,\text{W12}} + W_{t,\text{W14}} + W_{t,\text{W24}}$$
$$+ W_{t,\text{W8}} + W_{t,\text{HSS8}} + W_{t,\text{HSS14}} + W_{t,\text{HSS16}}$$

$$= \frac{\begin{array}{c}3840 \text{ lbf} + 1368 \text{ lbf} + 3744 \text{ lbf} \\ + 2496 \text{ lbf} + 12{,}096 \text{ lbf} + 2520 \text{ lbf} \\ + 3871 \text{ lbf} + 3821 \text{ lbf} + 3556 \text{ lbf}\end{array}}{2000 \ \frac{\text{lbf}}{\text{ton}}}$$

$$= 18.66 \text{ tons} \quad (19 \text{ tons})$$

The answer is (D).

Author Commentary

Quantity takeoffs for steel are often calculated based on tonnage of steel. Weights for each of these members can be found in the AISC *Steel Construction Manual*.

47. Calculate the additional material cost due to material tax.

$$C_{m,\text{tax}} = (\$750{,}000)(0.05) = \$37{,}500$$

Calculate the additional labor cost due to labor tax.

$$C_{l,\text{tax}} = (\$450{,}000)(0.18) = \$81{,}000$$

Determine the additional cost for bid bonds.

$$C_b = (0.12)(C_t + C_{m,\text{tax}} + C_{l,\text{tax}})$$
$$= (0.12)(\$2{,}750{,}000 + \$37{,}500 + \$81{,}000)$$
$$= \$344{,}220$$

Determine the additional cost due to profit.

$$C_p = (0.1)(C_t + C_{m,\text{tax}} + C_{l,\text{tax}} + C_b)$$
$$= (0.1)\left(\begin{array}{c}\$2{,}750{,}000 + \$37{,}500 \\ + \$81{,}000 + \$344{,}200\end{array}\right)$$
$$= \$321{,}270$$

Calculate the final bid.

$$\text{final bid} = C_t + C_{m,\text{tax}} + C_{l,\text{tax}} + C_b + C_p$$
$$= \$2{,}750{,}000 + \$37{,}500 + \$81{,}000$$
$$+ \$344{,}200 + \$321{,}270$$
$$= \$3{,}533{,}970 \quad (\$3{,}530{,}000)$$

The answer is (C).

Author Commentary

Be careful when adjusting costs. Some taxes, such as material tax and labor tax, are specific to certain items. Other adjustments, such as bid bonding percentages, are applied to total costs.

48. Determine the length of horizontal framing to be constructed for each wall.

$$L_{E-W} = 150 \text{ ft} + 5 \text{ ft} + 100 \text{ ft} = 255 \text{ ft}$$
$$L_{N-S} = 50 \text{ ft} + 5 \text{ ft} + 100 \text{ ft} + 100 \text{ ft} = 255 \text{ ft}$$
$$L_{\text{angled}} = \sqrt{(45 \text{ ft})^2 + (45 \text{ ft})^2}$$
$$= 63.64 \text{ ft}$$
$$L_t = 255 \text{ ft} + 255 \text{ ft} + 63.64 \text{ ft}$$
$$= 537.64 \text{ ft}$$

The total length of horizontal framing to be constructed is

$$L_{\text{horiz}} = L_t N_{\text{plates}} = \left(\frac{573.64 \text{ ft}}{\text{plate}}\right)(3 \text{ plates})$$
$$= 1720.92 \text{ ft}$$

Determine the total number of studs (vertical framing) needed.

$$N_s = \frac{(573.64 \text{ ft})\left(12 \frac{\text{in}}{\text{ft}}\right)}{16 \frac{\text{in}}{\text{stud}}}$$

$$= 430.23 \text{ studs} \quad (431 \text{ studs})$$

While the nominal dimensions of the boards are 2 in × 6 in, the actual dimensions are 1.5 in × 5.5 in. The header and footer are each 1.5 in tall. Calculate the height of the studs.

$$h_s = h_w - h_{\text{plates}} = 10 \text{ ft} - \frac{1.5 \text{ in} + 1.5 \text{ in} + 1.5 \text{ in}}{12 \frac{\text{in}}{\text{ft}}}$$

$$= 9.625 \text{ ft}$$

Determine the total length of studs (vertical framing) needed.

$$L_s = (431 \text{ studs})(9.625 \text{ ft})$$
$$= 4148.375 \text{ ft}$$

Calculate the total length of wood wasted.

$$L_{t,\text{waste}} = (L_{\text{horiz}} + L_s)\text{WF}$$
$$= (1720.92 \text{ ft} + 4148.375 \text{ ft})(0.05)$$
$$= 293.46 \text{ ft}$$

The total length of wood needed is

$$L_{t,\text{needed}} = L_{\text{horiz}} + L_s + L_{t,\text{waste}}$$
$$= 1720.92 \text{ ft} + 4148.375 \text{ ft} + 293.46 \text{ ft}$$
$$= 6162.755 \text{ ft} \quad (6200 \text{ ft})$$

The answer is (D).

Author Commentary

💣 When calculating the length of framing, remember to include the interior walls, as well as the header and footer. Also remember to deduct the header and footer heights from the total wall height.

49. Determine the cost of each project after VE.

$$C_{\text{VE},1} = \$3{,}250{,}000 - \$250{,}000 = \$3{,}000{,}000$$
$$C_{\text{VE},2} = \$6{,}750{,}000 - \$750{,}000 = \$6{,}000{,}000$$

Using the uniform series equivalence method, determine the payback period of each project.

For project 1 with VE,

$$P = A\left(\frac{(1+i)^n - 1}{i(1+i)^n}\right)$$

$$\$3{,}000{,}000 = (\$750{,}000)\left(\frac{(1+0.15)^n - 1}{(0.15)(1+0.15)^n}\right)$$

$$4 = \frac{(1+0.15)^n - 1}{(0.15)(1+0.15)^n}$$

$$(0.6)(1.15^n) = 1.15^n - 1$$
$$(0.4)(1.15^n) = 1$$
$$1.15^n = 2.5$$
$$n = \frac{\ln 2.5}{\ln 1.15} = 6.6 \text{ yr} \quad [>5 \text{ yr}]$$

For project 2 with VE,

$$P = A\left(\frac{(1+i)^n - 1}{i(1+i)^n}\right)$$

$$\$6{,}000{,}000 = (\$1{,}900{,}000)\left(\frac{(1+0.15)^n - 1}{(0.15)(1+0.15)^n}\right)$$

$$3.16 = \frac{(1+0.15)^n - 1}{(0.15)(1+0.15)^n}$$

$$(0.474)(1.15^n) = 1.15^n - 1$$
$$(0.526)(1.15^n) = 1$$
$$1.15^n = 1.90$$
$$n = \frac{\ln 1.90}{\ln 1.15} = 4.59 \text{ yr} \quad [<5 \text{ yr}]$$

Only project 2 yields a payback period of fewer than five years after VE.

The answer is (C).

Author Commentary

Many owners have a project approval process that considers the payback period of the project.

⏱ Instead of determining the exact payback period, this problem could be solved by calculating the payback period of both options with an n value of 5. If the result yielded a number higher than the initial investment, the payback period would be shorter than 5 years.

50. Calculate the total benefit of the project.

$$B = \$500{,}000{,}000$$

Calculate the total cost of the project.

$$C = \$325{,}000{,}000 + \$10{,}000{,}000 - \$40{,}000{,}000$$
$$= \$295{,}000{,}000$$

Determine the benefit-cost ratio.

$$B/C = \frac{\$500{,}000{,}000}{\$295{,}000{,}000} = 1.695 \quad (1.7)$$

The answer is (D).

Author Commentary

Engineering economics includes not only the time value of money, but also the analysis of benefit-cost ratios. A project with a benefit-cost ratio greater than or equal to 1.0 is considered to be acceptable.

💣 Read the problem carefully. Do not be distracted by extraneous information. Useful life and annual depreciation are not used to solve this problem.

51. Use a cash flow equivalent factor table to find the cash flow equivalent factor for 10 years at a 5% interest rate.

Calculate the present worth.

$$P = -\$213{,}000 - (\$600 + \$700 + \$1300)$$
$$\times (P/A, 5\%, 10) + (\$50{,}000)(P/F, 5\%, 10)$$
$$= -\$213{,}000 - (\$600 + \$700 + \$1300)(7.7217)$$
$$+ (\$50{,}000)(0.6139)$$
$$= -\$202{,}380 \quad (-\$202{,}000)$$

A negative present worth indicates a cost to the owner.

The answer is (B).

Alternative Solution

Convert the annuity and future value into present values, and calculate the present worth.

$$P = -\$213{,}000 - (\$600 + \$700 + \$1300)$$
$$\times (P/A, 5\%, 10) + (\$50{,}000)(P/F, 5\%, 10)$$
$$= -\$213{,}000 - (\$600 + \$700 + \$1300)$$
$$\times \left(\frac{1 - \frac{1}{(1+0.05)^{10}}}{0.05} \right) + (\$50{,}000)\left(\frac{1}{(1+0.05)^{10}} \right)$$
$$= -\$202{,}380 \quad (-\$202{,}000)$$

A negative present worth indicates a cost to the owner.

The answer is (B).

Author Commentary

Present costs, future costs, and annual costs can all be converted between each other with a fairly simple set of equations based on interest rates and compounding periods. Factor tables offer a faster way to solve the problem. Often times, interest in engineering economics problems is calculated monthly or quarterly, so be careful when calculating values with these types of interest. Remember that negative values indicate a cost to the owner, while positive numbers indicate a gain.

🕐 Tab the factor tables in a reference manual for easy access during the exam.

💣 Be sure to use the correct factor table and n value.

52. Draw the edges of the building on the chart, then follow the 140 ft boom curve to where it intersects with the 60° pitch line. Draw a vertical line from this intersection to the bottom of the chart. Check for vertical spacing requirements between the roof of the building and the boom of the crane (see *Illustration for Solution 52*).

The minimum distance from the roof of the building to the boom is 25 ft, which is greater than the 15 ft required. Therefore, the maximum operating radius is approximately 72 ft (70 ft).

The answer is (B).

Author Commentary

Crane operations depend on many different factors, including crane capacity, range, weather, and site conditions. When planning for major lifts on a jobsite, it is important to take all of these factors into consideration.

53. Complete the tipping diagram as shown.

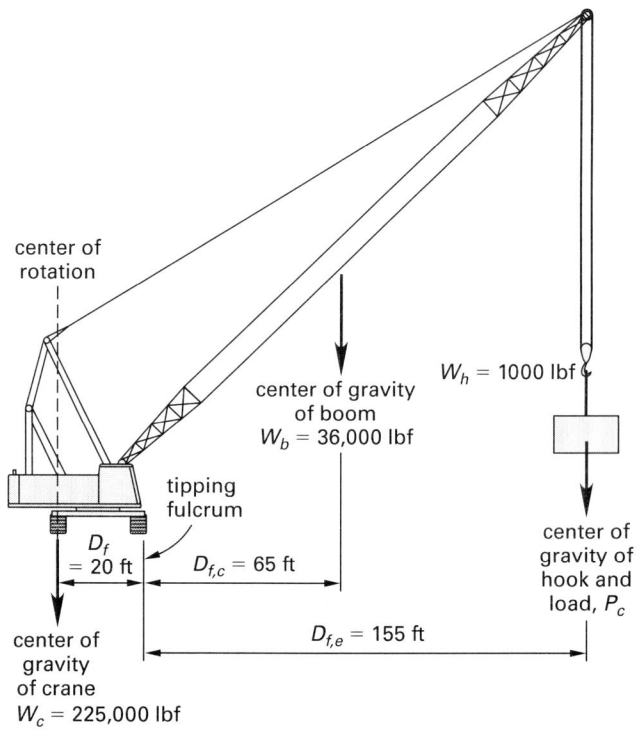

Illustration for Solution 52

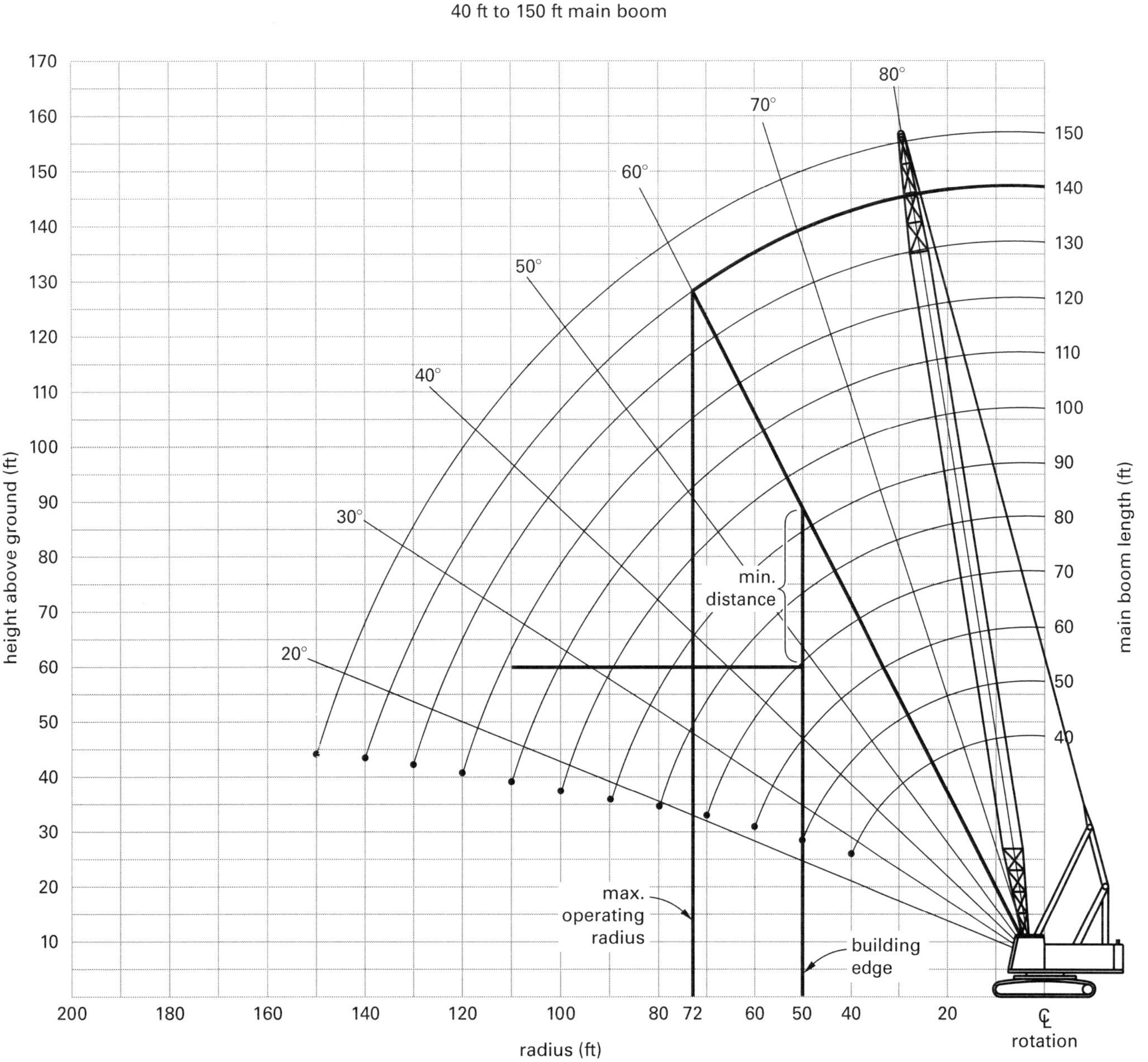

40 ft to 150 ft main boom

Calculate the tipping load.

$$P_c = \frac{W_c(D_{g,\text{CL}} + D_f) - W_b D_{f,c}}{D_{f,e}} - W_h$$

$$= \frac{(225{,}000 \text{ lbf})(0 \text{ ft} + 20 \text{ ft})}{155 \text{ ft}} - \frac{-(36{,}000 \text{ lbf})(65 \text{ ft})}{155 \text{ ft}} - 1000 \text{ lbf}$$

$$= 12{,}935.48 \text{ lbf} \quad (12{,}900 \text{ lbf})$$

The answer is (B).

Author Commentary

Mobile cranes can make picks without outriggers being set, but a tipping diagram must be completed.

💣 Remember to take moments about the tipping fulcrum.

💣 Don't forget to account for the weight of the hook.

54. From the hydraulic horsepower equation,

$$\text{WHP} = \frac{h_A Q(\text{SG})}{3956} = \frac{(125 \text{ ft})\left(300 \ \frac{\text{gal}}{\text{min}}\right)(1.0)}{3956} = 9.48 \text{ hp}$$

Adjust for 92% pump efficiency.

$$\text{WHP}_{\text{adj}} = \frac{9.48 \text{ hp}}{0.92} = 10.30 \text{ hp} \quad (10.3 \text{ hp})$$

The answer is (C).

Alternative Method

Determine the horsepower needed to pump the water at 300 gal/min.

$$\text{WHP} = \frac{\left(300 \, \frac{\text{gal}}{\text{min}}\right)\left(8.33 \, \frac{\text{lbf}}{\text{gal}}\right)(125 \text{ ft})}{33{,}000 \, \frac{\text{ft-lbf}}{\frac{\text{min}}{\text{hp}}}} = 9.47 \text{ hp}$$

Adjust for a 92% pump efficiency.

$$\text{WHP}_{\text{adj}} = \frac{9.47 \text{ hp}}{0.92} = 10.29 \text{ hp} \quad (10.3 \text{ hp})$$

The answer is (C).

Author Commentary

☉ Using the alternative method will be quicker during the exam.

55. Determine the fleet match for one truck. The fleet match is the cycle time of the excavator compared to the cycle time of one truck.

$$\text{fleet match}_{\text{no. trucks}} = \left(\frac{t_{\text{cycle,excavator}}}{t_{\text{cycle,truck}}}\right) N_{\text{trucks}}$$

$$\text{fleet match}_{1 \text{ truck}} = \left(\frac{3 \text{ min}}{15 \text{ min}}\right)(1) = 0.2$$

Bunching is the stacking of trucks at the excavator (loader). When bunching is at a maximum, efficiency is at a minimum. From the graph, the maximum bunching factor at a fleet match of 0.2 is 97% (0.97).

Determine the actual productivity of one truck.

$$P_{1 \text{ truck}} = P_{\text{ideal}}(\text{bunching factor})$$
$$= \left(115 \, \frac{\text{LCY}}{\text{hr}}\right)(0.97)$$
$$= 111.55 \text{ LCY/hr}$$

Repeat these calculations until the actual productivity is greater than the maximum productivity of the excavator, 550 LCY/hr.

For two trucks,

$$\text{fleet match}_{2 \text{ trucks}} = (0.2)(2) = 0.4$$
$$P_{2 \text{ trucks}} = \left(230 \, \frac{\text{LCY}}{\text{hr}}\right)(0.95)$$
$$= 218.50 \text{ LCY/hr} \quad (< 550 \text{ LCY/hr})$$

For three trucks,

$$\text{fleet match}_{3 \text{ trucks}} = (0.2)(3) = 0.6$$
$$P_{3 \text{ trucks}} = \left(345 \, \frac{\text{LCY}}{\text{hr}}\right)(0.90)$$
$$= 310.10 \text{ LCY/hr} \quad (< 550 \text{ LCY/hr})$$

For four trucks,

$$\text{fleet match}_{4 \text{ trucks}} = (0.2)(4) = 0.8$$
$$P_{4 \text{ trucks}} = \left(460 \, \frac{\text{LCY}}{\text{hr}}\right)(0.85)$$
$$= 391.00 \text{ LCY/hr} \quad (< 550 \text{ LCY/hr})$$

For five trucks,

$$\text{fleet match}_{5 \text{ trucks}} = (0.2)(5) = 1.0$$
$$P_{5 \text{ trucks}} = \left(575 \, \frac{\text{LCY}}{\text{hr}}\right)(0.80)$$
$$= 460.00 \text{ LCY/hr} \quad (< 550 \text{ LCY/hr})$$

For six trucks,

$$\text{fleet match}_{6 \text{ trucks}} = (0.2)(6) = 1.2$$
$$P_{6 \text{ trucks}} = \left(690 \, \frac{\text{LCY}}{\text{hr}}\right)(0.87)$$
$$= 600.30 \text{ LCY/hr} \quad (> 550 \text{ LCY/hr})$$

Six trucks are required to reach the maximum productivity of the excavator (600.3 LCY/hr > 550 LCY/hr).

The answer is (D).

Author Commentary

In situations where two types of equipment work together to accomplish a task, it is important that a balance in the productivity of the units be achieved. This is desirable so that one unit is not continually idle waiting for other unit to catch up. Bunching and queuing are an essential part of equipment production. Real life applications of this involve problems with traffic, restricted roadways, and other environmental factors. Without bunching and queuing analysis, the productivity of your jobsite will be extremely low and cause many inefficiencies.

56. Determine the amount of masonry required for the project.

$$Q_m = (3 \text{ walls})(100 \text{ ft})(16 \text{ ft}) - 300 \text{ ft}^2 = 4500 \text{ ft}^2$$

Calculate the productivity rate of the masonry operation that requires enclosures.

$$P_{\text{enclosures}} = \left(3 \ \frac{\text{enclosures}}{\text{day}}\right)(8 \text{ ft})(16 \text{ ft})$$
$$= 384 \text{ ft}^2/\text{day of masonry}$$

Find the durations of both operations.

$$D_{\text{enclosures}} = \frac{4500 \text{ ft}^2}{384 \ \frac{\text{ft}^2}{\text{day}}} = 11.7 \text{ days}$$

$$D_{\text{normal}} = \frac{4500 \text{ ft}^2}{750 \ \frac{\text{ft}^2}{\text{day}}} = 6 \text{ days}$$

Calculate the difference in durations.

$$D_{\text{diff}} = D_{\text{enclosures}} - D_{\text{normal}} = 11.7 \text{ days} - 6 \text{ days}$$
$$= 5.7 \text{ days} \quad (6 \text{ days})$$

The answer is (B).

Author Commentary

Adverse weather conditions can often affect the means and methods that are used on a construction site. There is often a time versus cost trade off and productivity analysis needed in order to decide which construction method is the best.

57. Determine the runoff prior to redevelopment (current conditions).

The combined runoff coefficient for the site is

$$C_{\text{pre}} = \frac{(4 \text{ ac})(0.2) + (2 \text{ ac})(0.9) + (1.5 \text{ ac})(0.85)}{7.5 \text{ ac}}$$
$$= 0.5167$$

Use the rational formula to find the rate of runoff. Although the calculation of Q will result in units of ac-in/hr, Q is taken as being in ft^3/sec, since the conversion factor between the two units is 1.008.

$$Q_{\text{pre}} = CiA_d = (0.5167)\left(2 \ \frac{\text{in}}{\text{hr}}\right)(7.5 \text{ ac})$$
$$= 7.7505 \text{ ac-in/hr} \quad (7.75 \text{ ft}^3/\text{sec})$$

Determine the runoff after redevelopment (planned conditions).

The combined runoff coefficient for the site is

$$C_{\text{post}} = \frac{\begin{array}{c}(2 \text{ ac})(0.2) + (1.5 \text{ ac})(0.05) + (1 \text{ ac})(0.9) \\ + (1 \text{ ac})(0.60) + (2 \text{ ac})(0.85)\end{array}}{7.5 \text{ ac}}$$
$$= 0.49$$

From the rational formula, the rate of runoff is

$$Q_{\text{post}} = CiA_d = (0.49)\left(2 \ \frac{\text{in}}{\text{hr}}\right)(7.5 \text{ ac})$$
$$= 7.35 \text{ ac-in/hr} \quad (7.35 \text{ ft}^3/\text{sec})$$

The answer is (B).

Author Commentary

The rational formula can be found in many reference books, including the United States Environmental Protection Agency's (USEPA's) 1992 *Stormwater Management for Construction Activities: Developing Pollution Prevention Plans and Best Management Practices* (EPA 832-R-92-005).

58. Create a critical path method (CPM) precedence diagram, and find all activities with zero total float (TF) (see *Illustration for Solution 58*).

Organize activities and perform a forward and backward pass. Once these steps are accomplished, identify the total float for all activities.

To perform a forward pass, determine the earliest start (ES) and earliest finish (EF) of each activity. For activity A, the ES value is 0 days because it is the first activity. The earliest finish for activity A is

$$\text{EF}_A = \text{ES}_A + D_A = 0 \text{ days} + 3 \text{ days}$$
$$= 3 \text{ days}$$

The ES value of activity B is 3 days because the EF value of its only predecessor, activity A, is 3 days. The earliest finish of activity B is

$$\text{EF}_B = \text{ES}_B + D_B = 3 \text{ days} + 5 \text{ days}$$
$$= 8 \text{ days}$$

Activities C and D each have only one predecessor, making their earliest start values easy to determine.

For activity C,

$$\text{ES}_C = \text{EF}_A = 3 \text{ days}$$
$$\text{EF}_C = \text{ES}_C + D_C = 3 \text{ days} + 7 \text{ days}$$
$$= 10 \text{ days}$$

For activity D,

$$ES_D = EF_A = 3 \text{ days}$$
$$EF_D = ES_D + D_D = 3 \text{ days} + 8 \text{ days}$$
$$= 11 \text{ days}$$

When two or more activities precede a given activity, the largest EF value should be used to determine the ES of the activity. This can be seen with activity E and activity F.

The ES of activity E is 17 days, the largest EF of its two predecessor activities, activity B and activity C. The earliest finish of activity E is

$$EF_E = ES_E + D_E = 10 \text{ days} + 4 \text{ days}$$
$$= 14 \text{ days}$$

The ES of activity F is 11 days, the largest EF of its two predecessor activities, activity C and activity D. The earliest finish of activity F is

$$EF_F = ES_F + D_F = 11 \text{ days} + 6 \text{ days}$$
$$= 17 \text{ days}$$

Activity G has only one predecessor, making its earliest start value easy to determine.

$$ES_G = EF_D = 11 \text{ days}$$
$$EF_G = ES_G + D_G$$
$$= 11 \text{ days} + 9 \text{ days}$$
$$= 20 \text{ days}$$

The ES of activity H is 17 days, the largest EF of its two predecessor activities, activity E and activity F. The earliest finish of activity H is

$$EF_H = ES_H + D_H = 17 \text{ days} + 5 \text{ days}$$
$$= 22 \text{ days}$$

Activity I has only one predecessor, making its earliest start value easy to determine.

$$ES_I = EF_F = 17 \text{ days}$$
$$EF_I = ES_I + D_I = 17 \text{ days} + 6 \text{ days}$$
$$= 23 \text{ days}$$

The ES of activity J is 23 days, the largest EF of its three predecessor activities, activity G, actitivy H, and activity I. The earliest finish of activity J is

$$EF_J = ES_J + D_J = 23 \text{ days} + 2 \text{ days}$$
$$= 25 \text{ days}$$

Since activity J is the last activity, the total duration of the project is 25 days.

Complete a backward pass to find the latest finish (LF) and latest start (LS) of each activity. Since activity J is the final activity, its EF value is equal to its LF value. The LF of activity J is 25 days. The LS of activity J is

$$LS_J = LF_J - D_J = 25 \text{ days} - 2 \text{ days}$$
$$= 23 \text{ days}$$

Repeat these steps for the other activities. Activities I, H, and G each have only one successor, making it easy to find their LF values.

Illustration for Solution 58

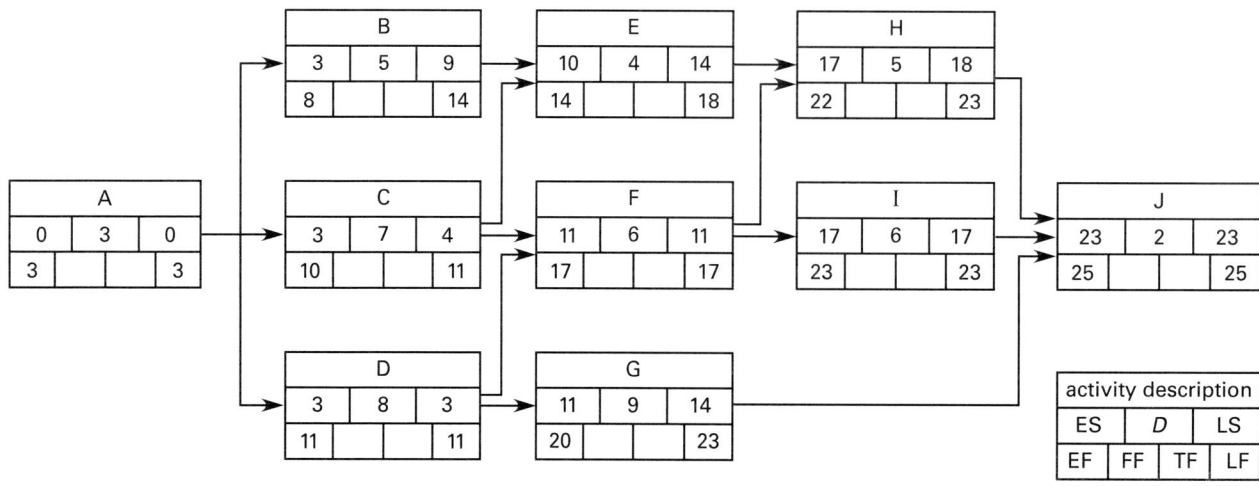

For activity I,

$$LF_I = LS_J = 23 \text{ days}$$
$$LS_I = LF_I - D_I = 23 \text{ days} - 6 \text{ days}$$
$$= 17 \text{ days}$$

For activity H,

$$LF_H = LS_J = 23 \text{ days}$$
$$LS_H = LF_H - D_H = 23 \text{ days} - 5 \text{ days}$$
$$= 18 \text{ days}$$

For activity G,

$$LF_G = LS_J = 23 \text{ days}$$
$$LS_G = LF_G - D_G = 23 \text{ days} - 9 \text{ days}$$
$$= 14 \text{ days}$$

The LF of activity F is 17 days, the smallest LS of its two successor activities, activity H and activity I. The latest start of activity F is

$$LS_F = LF_F - D_F = 17 \text{ days} - 6 \text{ days}$$
$$= 11 \text{ days}$$

Activity E has only one successor, making it easy to find its LF value.

$$LF_E = LS_H = 18 \text{ days}$$
$$LS_E = LF_E - D_E = 18 \text{ days} - 4 \text{ days}$$
$$= 14 \text{ days}$$

The LF of activity D is 11 days, the smallest LS of its two successor activities, activity F and activity G. The latest start of activity D is

$$LS_D = LF_D - D_D = 11 \text{ days} - 8 \text{ days}$$
$$= 3 \text{ days}$$

The LF of activity C is 11 days, the smallest LS of its two successor activities, activity E and activity F. The latest start of activity C is

$$LS_C = LF_C - D_C$$
$$= 11 \text{ days} - 7 \text{ days}$$
$$= 4 \text{ days}$$

Activity B has only one successor, making it easy to find its LF value.

$$LF_B = LS_E = 14 \text{ days}$$
$$LS_B = LF_B - D_B = 14 \text{ days} - 5 \text{ days}$$
$$= 9 \text{ days}$$

The LF of activity A is 3 days, the smallest LS of its three successor activities, activity B, activity C, and activity D. The latest start of activity A is

$$LS_A = LF_A - D_A = 3 \text{ days} - 3 \text{ days}$$
$$= 0$$

Determine the total float (TF) of each activity.

For activity A,

$$TF_A = LS_A - ES_A = 0 \text{ days} - 0 \text{ days}$$
$$= 0$$

For activity B,

$$TF_B = LS_B - ES_B = 9 \text{ days} - 3 \text{ days}$$
$$= 6 \text{ days}$$

For activity C,

$$TF_C = LS_C - ES_C = 4 \text{ days} - 3 \text{ days}$$
$$= 1 \text{ day}$$

For activity D,

$$TF_D = LS_D - ES_D = 3 \text{ days} - 3 \text{ days}$$
$$= 0$$

For activity E,

$$TF_E = LS_E - ES_E = 14 \text{ days} - 10 \text{ days}$$
$$= 4 \text{ days}$$

For activity F,

$$TF_F = LS_F - ES_F = 11 \text{ days} - 11 \text{ days}$$
$$= 0$$

For activity G,

$$TF_G = LS_G - ES_G = 14 \text{ days} - 11 \text{ days}$$
$$= 3 \text{ days}$$

For activity H,

$$TF_H = LS_H - ES_H = 18 \text{ days} - 17 \text{ days}$$
$$= 1 \text{ day}$$

For activity I,

$$TF_I = LS_I - ES_I = 17 \text{ days} - 17 \text{ days}$$
$$= 0$$

For activity J,

$$TF_J = LS_J - ES_J = 23 \text{ days} - 23 \text{ days}$$
$$= 0$$

Activities that have zero total float are on the critical path. Zero float means the activity cannot be delayed without impacting the final duration of the project.

Activities A, D, F, I, and J are on the critical path.

The answer is (D).

59. The term *predecessor* refers to activities occurring before another activity. The term *successor* refers to activities occurring after another activity. Using the values given in the table, the diagram for option D is correct.

The answer is (D).

Author Commentary

Being able to produce a logic diagram such as the one in this problem can provide the owner and the contractor with a clear understanding of how a construction project will be executed. Although not everyone involved on a project will be able to quickly understand successors and predecessors in construction activities, they are often able to read this type of diagram.

60. Complete a forward pass to determine the original completion date.

To perform a forward pass, determine the earliest start (ES) and earliest finish (EF) of each activity. For activity A, the ES value is 0 days because it is the first activity. The earliest finish for activity A is

$$EF_A = ES_A + D_A = 0 \text{ days} + 2 \text{ days}$$
$$= 2 \text{ days}$$

The ES value of activity B is 2 days, because the EF value of its only predecessor, activity A, is 2 days. The earliest finish of activity B is

$$EF_B = ES_B + D_B = 2 \text{ days} + 2 \text{ days}$$
$$= 4 \text{ days}$$

Repeat these steps for each subsequent activity. Activities D, C, E, G, H, and I each have only one predecessor, making their earliest start values easy to determine.

For activity D,

$$ES_D = EF_A = 2 \text{ days}$$
$$EF_D = ES_D + D_D = 2 \text{ days} + 5 \text{ days}$$
$$= 7 \text{ days}$$

For activity C,

$$ES_C = EF_D = 7 \text{ days}$$
$$EF_C = ES_C + D_C = 7 \text{ days} + 4 \text{ days}$$
$$= 11 \text{ days}$$

For activity E,

$$ES_E = EF_B = 4 \text{ days}$$
$$EF_E = ES_E + D_E = 4 \text{ days} + 4 \text{ days}$$
$$= 8 \text{ days}$$

For activity G,

$$ES_G = EF_E = 8 \text{ days}$$
$$EF_G = ES_G + D_G = 8 \text{ days} + 4 \text{ days}$$
$$= 12 \text{ days}$$

For activity H,

$$ES_H = EF_C = 11 \text{ days}$$
$$EF_H = ES_H + D_H = 11 \text{ days} + 5 \text{ days}$$
$$= 16 \text{ days}$$

For activity I,

$$ES_I = EF_G = 12 \text{ days}$$
$$EF_I = ES_I + D_I = 12 \text{ days} + 8 \text{ days}$$
$$= 20 \text{ days}$$

The ES of activity F is 16 days, the largest EF of its two predecessor activities, activity D and activity H. The earliest finish of activity F is

$$EF_F = ES_F + D_F = 16 \text{ days} + 7 \text{ days}$$
$$= 23 \text{ days}$$

The ES of activity J is 23 days, the largest EF of its two predecessor activities, activity F and activity I. The earliest finish of activity J is

$$EF_J = ES_J + D_J = 23 \text{ days} + 3 \text{ days}$$
$$= 26 \text{ days}$$

The original schedule will take 26 days to complete (see *Illustration for Solution 60(a)*).

Add activity K into the sequence and complete a forward pass again to determine the new completion date.

For activity A, the ES value is 0 days because it is the first activity. The earliest finish for activity A is

$$EF_A = ES_A + D_A = 0 \text{ days} + 2 \text{ days} = 2 \text{ days}$$

The ES value of activity B is 2 days, because the EF value of its only predecessor, activity A, is 2 days. The earliest finish for activity B is

$$EF_B = ES_B + D_B = 2 \text{ days} + 2 \text{ days} = 4 \text{ days}$$

Repeat these steps for each subsequent activity. Activities D, C, E, G, H, and I each have only predecessor, making their earliest start values easy to determine.

For activity D,

$$ES_D = EF_A = 2 \text{ days}$$
$$EF_D = ES_D + D_D = 2 \text{ days} + 5 \text{ days}$$
$$= 7 \text{ days}$$

For activity C,

$$ES_C = EF_D = 7 \text{ days}$$
$$EF_C = ES_C + D_C = 7 \text{ days} + 4 \text{ days}$$
$$= 11 \text{ days}$$

For activity E,

$$ES_E = EF_B = 4 \text{ days}$$
$$EF_E = ES_E + D_E = 4 \text{ days} + 4 \text{ days}$$
$$= 8 \text{ days}$$

For activity G,

$$ES_G = EF_E = 8 \text{ days}$$
$$EF_G = ES_G + D_G = 8 \text{ days} + 4 \text{ days}$$
$$= 12 \text{ days}$$

For activity H,

$$ES_H = EF_C = 11 \text{ days}$$
$$EF_H = ES_H + D_H = 11 \text{ days} + 5 \text{ days}$$
$$= 16 \text{ days}$$

For activity I,

$$ES_I = EF_G = 12 \text{ days}$$
$$EF_I = ES_I + D_I = 12 \text{ days} + 8 \text{ days}$$
$$= 20 \text{ days}$$

The ES of activity K is 11 days, the largest EF of its three predecessor activities, activity C, activity D, and activity E. The earliest finish of activity K is

$$EF_K = ES_K + D_K = 11 \text{ days} + 6 \text{ days}$$
$$= 17 \text{ days}$$

The ES of activity F is 17 days, the largest EF of its two predecessor activities, activity K and activity H. The earliest finish of activity F is

$$EF_F = ES_F + D_F$$
$$= 17 \text{ days} + 7 \text{ days}$$
$$= 24 \text{ days}$$

Illustration for Solution 60(a)

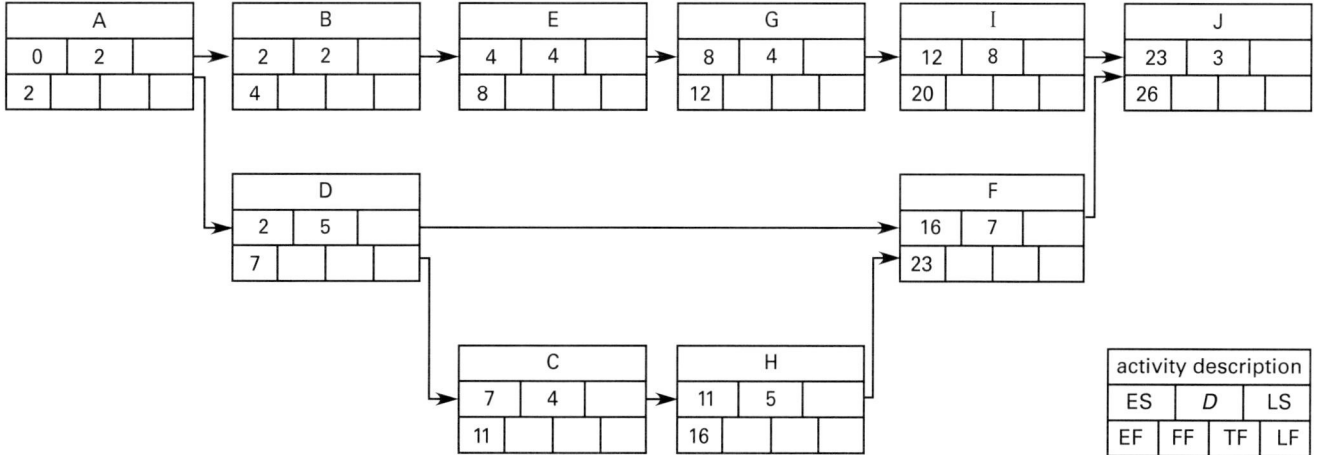

The ES of activity J is 23 days, the largest EF of its two predecessor activities, activity F and activity I. The earliest finish of activity J is

$$EF_J = ES_J + D_J = 24 \text{ days} + 3 \text{ days}$$
$$= 27 \text{ days}$$

The revised schedule (*see Illustration for Solution 60(b)*) will take 27 days to complete, so one additional day will be needed when activity K is added into the sequence.

The answer is (B).

Author Commentary

The addition of an activity on a construction project does not simply delay the project by the duration of the new activity. Rather, it affects the sequence of all activities. If a new activity does not appear on the critical path, an extension may not be allowed by an owner.

61. A stacking of trades occurs when contractors must work in a limited space with other contractors. Adjust the crew's productivity rate to account for 75% efficiency.

$$P_{\text{crew,adj}} = \left(\frac{60 \frac{\text{min}}{\text{hr}}}{25 \frac{\text{min}}{\text{fixture}}}\right)(0.75) = 1.8 \text{ fixtures/hr}$$

Divide the number of fixtures by the adjusted crew rate to determine the total number of hours it will take to install the fixtures.

$$N_{\text{hr}} = \frac{45 \text{ fixtures}}{1.8 \frac{\text{fixtures}}{\text{hr}}} = 25 \text{ hr}$$

The answer is (D).

Illustration for Solution 60(b)

Author Commentary

Crew productivity is affected by many other factors besides stacking of trades. Other items that may affect productivity include number of crews, level of experience of workers, weather, safety, worker morale, repeatable tasks, and the difficulty of the work. Estimators and schedulers must be diligent when planning a job and know the details of the plan and sequence of tasks.

62. Determine how many days it will take the electricians to install the conduit.

$$N_{\text{days}} = \frac{15{,}500 \text{ ft}}{\left(120 \frac{\text{ft}}{\text{hr}}\right)\left(8 \frac{\text{hr}}{\text{day}}\right)}$$
$$= 16.15 \text{ days} \quad [>15 \text{ days; late}]$$

Determine the cost per day of the crew.

$$C_{\text{crew}} = (C_{\text{apprentice}} + 2C_{\text{journeymen}} + C_{\text{equipment}})N_{\text{hr}}$$
$$= \left(20 \frac{\$}{\text{hr}} + (2)\left(35 \frac{\$}{\text{hr}}\right) + 5 \frac{\$}{\text{hr}}\right)\left(8 \frac{\text{hr}}{\text{day}}\right)$$
$$= \$760/\text{day}$$

Calculate the total labor cost.

$$C_t = (16.15 \text{ days})\left(760 \frac{\$}{\text{day}}\right)$$
$$= \$12{,}274 \quad [>\$11{,}500; \text{ over budget}]$$

The answer is (D).

Author Commentary

Determining project costs and durations on a jobsite is a necessity for a construction manager. Accurately determining costs and durations will allow contractors to make more money and complete projects on time.

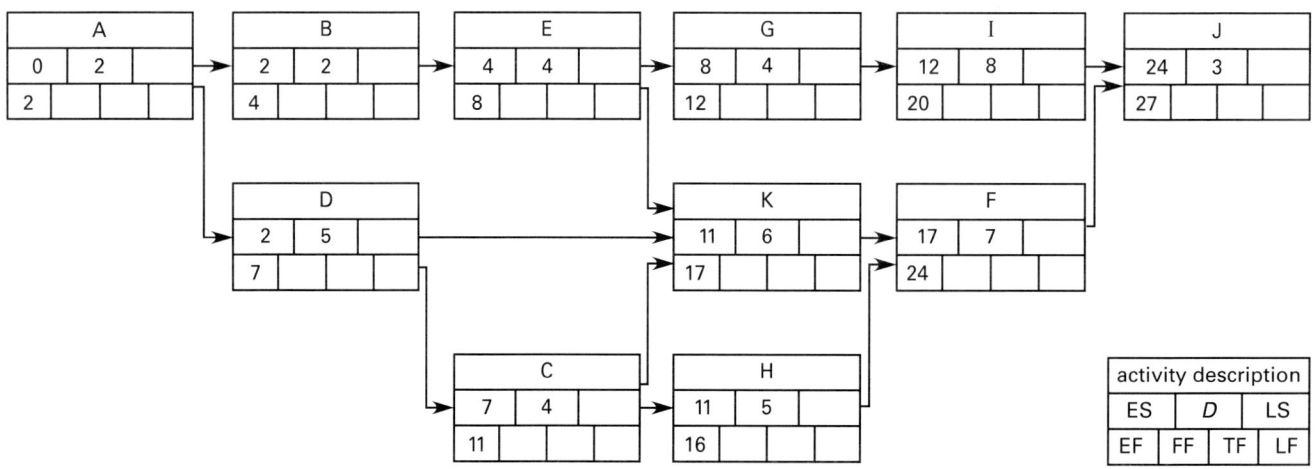

63. Graph the resources given in the table. The project considered most level is the one whose graph represents the most uniform distribution of resources (i.e., the fewest peaks and valleys).

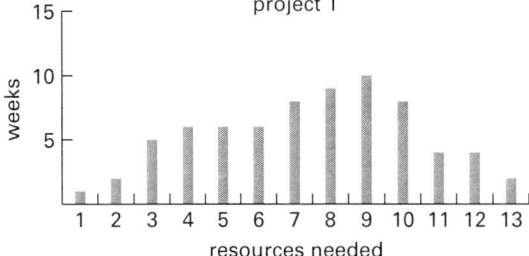

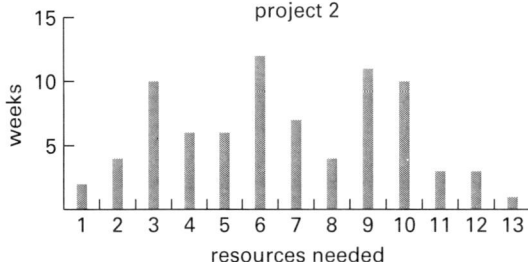

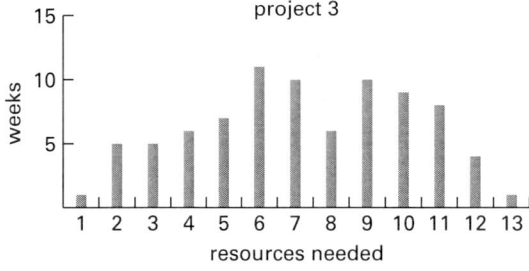

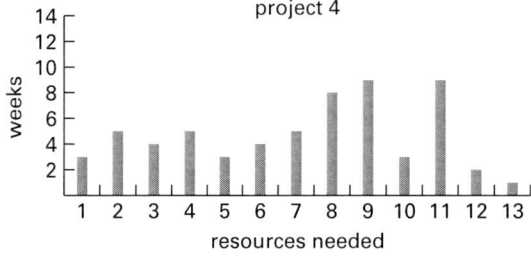

The answer is (A).

Author Commentary

Resource leveling is a scheduling tool to help managers balance the number of crews and the amount of equipment and stored materials on site. A schedule that demands a peak in resources followed by a dip in resources, makes it difficult for crews and equipment to be used efficiently. Sometimes, the activity sequence will cause peaks and dips to be unavoidable, but resources should be balanced throughout a project whenever possible.

64. Determine the hourly rate for the crane operation.

$$C_{\text{crane}} = 250 \ \frac{\$}{\text{hr}} + 75 \ \frac{\$}{\text{hr}} + 55 \ \frac{\$}{\text{hr}} + 55 \ \frac{\$}{\text{hr}}$$
$$= \$435/\text{hr}$$

Determine the hourly rate for the helicopter operation.

$$C_{\text{helicopter}} = 1000 \ \frac{\$}{\text{hr}} + 300 \ \frac{\$}{\text{hr}} + 65 \ \frac{\$}{\text{hr}} + 65 \ \frac{\$}{\text{hr}} + 40 \ \frac{\$}{\text{hr}}$$
$$= \$1470/\text{hr}$$

Determine the total cost of the crane operation.

$$C_t = N_{\text{hr}} C_{\text{crane}}$$
$$= \left(\frac{55 \text{ units}}{\frac{5 \text{ units}}{\text{hr}}} \right) \left(435 \ \frac{\$}{\text{hr}} \right)$$
$$= \$4785$$

Divide the total cost of the crane operation by the hourly rate of the helicopter operation.

$$N_{\text{hr,max}} = \frac{\$4785}{\frac{\$1470}{\text{hr}}}$$
$$= 3.26 \text{ hr} \quad (3 \text{ hr})$$

Check this by calculating the cost of operating the helicopter for three hours.

$$C_{\text{helicopter}} = \left(1470 \ \frac{\$}{\text{hr}} \right) (3 \text{ hr})$$
$$= \$4410 \quad [< \$4785, \text{ OK}]$$

The answer is (C).

Author Commentary

Analysis of a schedule often comes down to a time versus cost trade off. Sometimes, a method of construction may seem expensive, but after a time-cost analysis, the method may be found to be more efficient.

65. Find the volume of the trench that will be dug by the contractor.

$$V_{\text{trench}} = \frac{\left((4 \text{ ft})(4 \text{ ft}) + \frac{(2)(4 \text{ ft})(4 \text{ ft})}{2} \right)(300 \text{ ft})}{\left(3 \ \frac{\text{ft}}{\text{yd}} \right)^3}$$
$$= 356 \text{ yd}^3$$

Calculate the number of cycles needed to complete the work for the contractor.

$$N_{\text{cycles}} = \frac{356 \text{ yd}^3}{4 \frac{\text{yd}^3}{\text{cycle}}} = 89 \text{ cycles}$$

The duration of the activity for the contractor is

$$D = \frac{(89 \text{ cycles})\left(120 \frac{\text{sec}}{\text{cycle}}\right)}{3600 \frac{\text{sec}}{\text{hour}}} = 2.96 \text{ hr}$$

Find the volume of the trench that will be dug by the subcontractor.

$$V_{\text{trench}} = \frac{(3 \text{ ft})(4 \text{ ft})(300 \text{ ft})}{\left(3 \frac{\text{ft}}{\text{yd}}\right)^3} = 133 \text{ yd}^3$$

Calculate the number of cycles needed to complete the work for the subcontractor.

$$N_{\text{cycles}} = \frac{133 \text{ yd}^3}{2 \frac{\text{yd}^3}{\text{cycle}}} = 66.5 \text{ cycles}$$

The duration of the activity for the subcontractor is

$$D = \frac{(66.5 \text{ cycles})\left(55 \frac{\text{sec}}{\text{cycle}}\right)}{3600 \frac{\text{sec}}{\text{hr}}} = 1.02 \text{ hr}$$

Only the subcontractor would be able to complete the trench in less than 1.5 hr.

The answer is (B).

Author Commentary

Subcontractors in many cases have much more specialized equipment and very experienced operators in their specific fields. Contractors must explore these options in order to utilize the best subcontractors to perform the work economically.

66. IBC Sec. 2105.3.1 states that three prisms (at least 28 days old) must be cut from the masonry for each 5000 ft^2 of wall area. At least three prisms must be cut, regardless of the wall area.

Find the wall area of the building.

$$A = \begin{pmatrix} 200 \text{ ft} + 400 \text{ ft} + 100 \text{ ft} \\ + 100 \text{ ft} + 100 \text{ ft} + 300 \text{ ft} \end{pmatrix}(25 \text{ ft})$$

$$= 30{,}000 \text{ ft}^2$$

Therefore, the number of prisms required is

$$N_{\text{prisms}} = \frac{30{,}000 \text{ ft}^2}{\frac{5000 \text{ ft}^2}{3 \text{ prisms}}}$$

$$= 18 \text{ prisms}$$

The answer is (D).

Author Commentary

Contractors must provide materials that meet the specifications of their contracts. The prism test is a material testing method to measure the compressive strength of masonry and verify compliance of the masonry materials with the specifications.

- Use the index of the IBC to quickly find the quality assurance section for masonry.
- Make sure to calculate the area of the wall, not the length in order to get the correct calculation.

67. Refer to OSHA Sec. 1926.755(a) and Sec. 1926.755(b). OSHA Sec. 1926.755(a) states that columns must be anchored by a minimum of four anchor rods, be designed to resist a minimum eccentric gravity load of 300 lbf 18 in from the outer faces in each direction at the tops of the shafts, and be set on level finished floors adequate to transfer construction loads. OSHA Sec. 1926.755(b) states that anchor rods must not be repaired, replaced, or modified in the field without the approval of the project structural engineer, and that a senior project manager is not qualified to approve repairs, replacements, or field modifications.

The answer is (A).

Author Commentary

- Review OSHA Part 1926 prior to the exam and tab OSHA Sec. 1926.755 for easy access during the exam.
- It is important to understand the authority levels of those with titles such as project manager and project structural engineer.

68. IBC Sec. 1005.3.2 states that the capacities of means of egress components other than stairways are calculated using a capacity factor of 0.2 in/occupant.

$$w = N_{\text{occupants}}(\text{occupant factor})$$

$$= (600 \text{ occupants})\left(0.2 \frac{\text{in}}{\text{occupant}}\right)$$

$$= 120 \text{ in}$$

The number of openings required is

$$N_\text{openings} = \frac{120 \text{ in}}{36 \ \dfrac{\text{in}}{\text{opening}}}$$
$$= 3.33 \text{ openings} \quad (4 \text{ openings})$$

The answer is (C).

Author Commentary

When calculating numbers of openings for egress, all values must be rounded up to achieve the minimum requirements of the building code.

It should also be noted that if there are multiple means of egress, they must be sized so that if one is lost, it will not reduce the available capacity to less than 50% of the required capacity.

- 🕐 Become familiar with and tab the Means of Egress chapter in the IBC for easy access during the exam.
- 💣 It is important to consider all applicable codes in order to meet or exceed all requirements. For many problems, referencing IBC Sec. 1005 alone would not be enough to arrive at the correct solution.

69. Determine the weight of the cement using the absolute volume equation.

$$V_\text{absolute} = \frac{w}{(\text{SG})\gamma_\text{water}}$$
$$W_c = V_\text{absolute}(\text{SG})\gamma_w$$
$$= (10.15 \text{ ft}^3)(3.14)\left(62.4 \ \frac{\text{lbf}}{\text{ft}^3}\right)$$
$$= 1988.75 \text{ lbf}$$

Calculate the amount of water required for a water-cement ratio of 0.45.

$$W_w = \frac{(w\text{-}c)\,W_c}{\gamma_w}$$
$$= \frac{(0.45)(1988.75 \text{ lbf})\left(7.48 \ \dfrac{\text{gal}}{\text{ft}^3}\right)}{62.4 \ \dfrac{\text{lbf}}{\text{ft}^3}}$$
$$= 107.27 \text{ gal} \quad (110 \text{ gal})$$

The answer is (B).

Author Commentary

- 💣 The water-cement ratio is based on the weight of the water and cement only. Do not be distracted by the aggregate and air information provided.

70. Draw the free-body diagram for the bolt.

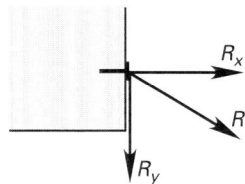

Draw the free-body diagram for the beam.

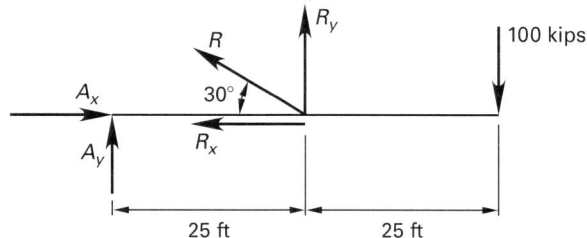

Determine the vertical reaction, R_y, at the center of the beam.

$$\Sigma M_A = (-100 \text{ kips})(50 \text{ ft}) + R_y(25 \text{ ft}) = 0$$
$$R_y = 200 \text{ kips}$$

Determine the reaction in the cable.

$$R = \frac{R_y}{\sin\theta} = \frac{200 \text{ kips}}{\sin 30°} = 400 \text{ kips}$$

The horizontal pull-out force on the bolt, R_x, is.

$$R_x = R\cos\theta = (400 \text{ kips})(\cos 30°)$$
$$= 346 \text{ kips} \quad (350 \text{ kips})$$

The answer is (C).

Author Commentary

- 🕐 It is unnecessary to solve for the support reactions at the left restraint of the beam. The force in the cable can be found by summing moments around the left end.
- 💣 Remember that sines and cosines are equal to the leg times the hypotenuse of the angle being solved for.

71. For concrete weighing less than 140 lbf/ft³, use ACI 347 Table 2.1 to find the equation for determining the unit weight coefficient, C_w.

$$C_w = 0.5\left(1 + \frac{W}{145 \ \dfrac{\text{lbf}}{\text{ft}}}\right) = (0.5)\left(1 + \frac{130 \ \dfrac{\text{lbf}}{\text{ft}}}{145 \ \dfrac{\text{lbf}}{\text{ft}}}\right)$$
$$= 0.948$$

Since the wall exceeds 14 ft, use ACI 347 Eq. 2.4. C_w and C_c are constants found in ACI 347 Table 2.1 and Table 2.2.

$$\rho_{\max} = C_w C_c \left(150 \, \frac{\text{lbf}}{\text{ft}^2} + \frac{43{,}400}{T} + \frac{2800R}{T} \right)$$

$$R = \left(\frac{T}{2800} \right) \left(\frac{\rho_{\max}}{C_w C_c} - 150 \, \frac{\text{lbf}}{\text{ft}^2} - \frac{43{,}400}{T} \right)$$

$$= \left(\frac{70°}{2800} \right) \left(\frac{1200 \, \frac{\text{lbf}}{\text{ft}^2}}{(0.948)(1.0)} - 150 \, \frac{\text{lbf}}{\text{ft}^2} - \frac{43{,}400}{70°} \right)$$

$$= 12.40 \text{ ft/hr} \quad (12 \text{ ft/hr})$$

The answer is (B).

Author Commentary

- Become familiar with ACI 347 and tab this equation for easy access during the exam.
- Be careful when rearranging the equation for rate of placement.
- Be sure to use the seventh edition of ACI 347, which has been adopted by NCEES. Earlier versions of ACI 347 do not acknowledge and apply the unit weight coefficient, C_w, and chemistry coefficient, C_c.

72. Use Rankine theory to find k_a, the coefficient of active earth pressure on the vertical wall face (principal plane).

On the right side,

$$k_{a,r} = \frac{1 - \sin \phi}{1 + \sin \phi} = \frac{1 - \sin 25°}{1 + \sin 25°} = 0.406$$

On the left side,

$$k_{a,l} = \frac{1 - \sin \phi}{1 + \sin \phi} = \frac{1 - \sin 35°}{1 + \sin 35°} = 0.271$$

Horizontal soil pressure is 0 at the soil surface. On the right side, it increases linearly to

$$p_{h,r} = k_a \gamma h = (0.406) \left(120 \, \frac{\text{lbf}}{\text{ft}^3} \right) (16 \text{ ft})$$

$$= 779.5 \text{ lbf/ft}^2$$

On the left side, it increases linearly to

$$p_{h,l} = k_a \gamma h = (0.271) \left(130 \, \frac{\text{lbf}}{\text{ft}^3} \right) (5 \text{ ft})$$

$$= 176.2 \text{ lbf/ft}^2$$

The horizontal resultant force on the right side is

$$R_{h,r} = \tfrac{1}{2} p_h h = \left(\tfrac{1}{2} \right) \left(779.5 \, \frac{\text{lbf}}{\text{ft}^2} \right) (16 \text{ ft})$$

$$= 6236 \text{ lbf/ft}$$

The horizontal resultant force on the left side is

$$R_{h,l} = \tfrac{1}{2} p_h h = \left(\tfrac{1}{2} \right) \left(176.2 \, \frac{\text{lbf}}{\text{ft}^2} \right) (5 \text{ ft})$$

$$= 440.5 \text{ lbf/ft}$$

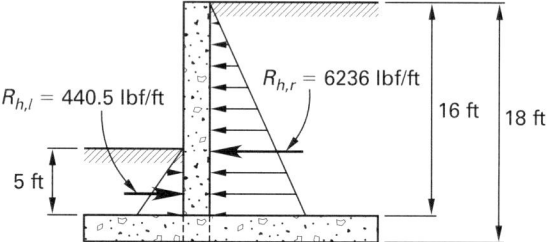

The net resultant horizontal force is

$$R_{h,\text{net}} = R_{h,r} - R_{h,l} = 6236 \, \frac{\text{lbf}}{\text{ft}} - 440.5 \, \frac{\text{lbf}}{\text{ft}}$$

$$= 5795.5 \text{ lbf/ft from right to left} \quad (5800 \text{ lbf/ft})$$

The answer is (A).

Author Commentary

Completing retaining wall problems may require knowing which direction the resultant force is being applied from (or to). The resultant force direction is from the larger force direction to the smaller force direction.

- When solving a retaining wall problem, be sure to use the depth of the soil to the top of the footing.

73. Per ACI 347 Chap. 5, four assumptions for the purposes of analysis of loads can be made. Options A, C, and D are identified as three of the four assumptions. The fourth assumption is that slabs interconnected by shores deflect equally when a new load is added, and they carry a share of the load in proportion to their relative stiffness.

Therefore, option B is an incorrect assumption.

The answer is (B).

Author Commentary

- Become familiar with ACI 347 and bring it as a reference during the exam.

74. According to the OSHA *Log of Work-Related Injuries and Illnesses* (Form 300), all injuries that require only the use of first aid are not considered recordable injuries or illnesses. Administration of a tetanus shot is considered to be first aid treatment.

The answer is (D).

Author Commentary

- Descriptions of OSHA recordable injuries or illnesses are not found in OSHA Part 1926. Bring OSHA Form 300 to the exam for easy reference.

75. OSHA Part 1926 Subpart E pertains to personal protective and lifesaving equipment. Per OSHA Sec. 1926.95 through Sec. 1926.107, options A, B, and D are true regarding personal protection and lifesaving equipment.

The answer is (C).

Author Commentary

- OSHA Part 1926 is a large document of over 600 pages. Although the first four pages provide a general table of contents, the starting page number of each section is not identified. Use tabs to mark frequently used sections to allow for easy access during the exam.

76. Options A, C, and D are given as requirements in OSHA Part 1926 Subpart L (Sec. 1926.451 to Sec. 1926.454 and applicable appendices).

The answer is (B).

Author Commentary

It is important to read the regulations carefully, including all subsections.

- OSHA Part 1926 is a large document of over 600 pages. Although the first four pages provide a general table of contents, the starting page number of each section is not identified. Use tabs to mark frequently used sections to allow for easy access during the exam.

77. Per OSHA Part 1926 Table B-1 Note 3, a registered professional engineer must design sloping or benching for excavations more than 20 ft deep.

The answer is (C).

Author Commentary

- Guidance for construction safety can be found in OSHA Part 1926. Tab Table B-1 for easy access during the exam.

78. To solve this problem, use a reference book such as *Formwork for Concrete* (see Codes and References) to find information needed, such as nail sizes, adjustment factors, and withdrawal load design values.

16 penny common nails are 3.5 in long and have a 0.344 in diameter head and a 0.162 diameter shank. The adjustment factor, C_M, for wood with 19% moisture content is 0.25. The withdrawal load design value for southern pine with the specified diameter nail is 50 lbf/in of penetration. The actual dimensions of a 2×4 beam are 1.5 in $\times$ 3.5 in.

Determine the length of penetration of the nails.

$$L_p = L_n - w_w = 3.5 \text{ in} - 1.5 \text{ in} = 2 \text{ in}$$

Calculate the allowable load per nail, taking into account the adjustment factor.

$$\begin{aligned} W_a &= w L_p C_M \\ &= \left(50 \; \frac{\text{lbf}}{\text{in of penetration}}\right) \\ &\quad \times \left(2 \; \frac{\text{in of penetration}}{\text{nail}}\right)(0.25) \\ &= 25 \text{ lbf/nail} \end{aligned}$$

The allowable load for the complete assembly is

$$W_{a,t} = \left(25 \; \frac{\text{lbf}}{\text{nail}}\right)(2 \text{ nails}) = 50 \text{ lbf}$$

The answer is (B).

Author Commentary

- Obtain a copy of a concrete reference book, such as *Formwork for Concrete* (see Codes and References). Become familiar with and tab all charts for easy reference during the exam.
- Multiplying the force by 2 is important, since the allowable load found in the second calculation is per nail, and not for a combination of both nails.

79. According to MUTCD Part 6, II and III are true.

The answer is (D).

Author Commentary

At longer distances, it is much more difficult to gauge how long it will take for traffic to completely clear out of the traveling lanes. Using a pilot car eliminates this problem and allows for more control over the jobsite.

80. First, determine the major division. Since less than 50% of the soil passes through the no. 200 sieve, the sample is classified as a coarse-grained soil. More than

50% of the soil passes through the no. 4 sieve, so the sample is further classified as a sandy soil.

Next, find the group. Since the percentage of soil passing through the no. 200 sieve is greater than 12%, the sample will be either SM or SC.

Calculate the plasticity index.

$$PI = LL - PL = 45 - 32$$
$$= 13$$

Refer to the plasticity chart for fine-grained soils, and plot the PI (13) and the LL (45). Since this point is below the A-line, the soil sample is classified as SM, a silty sand.

The answer is (A).

Author Commentary

🕐 Review the Unified Soil Classification System (USCS) chart (found in ASTM D2487) and have it on hand during the exam for quick reference.

💣 Read the requirements for determining the group carefully. For some soils, the PI values and plot point must both be met in order to classify the soil. In other cases, the PI value or plot point must be met in order to classify the soil.